Anglo-French Physical Acoustics Conference 2008

Journal of Physics: Conference Series Volume 195

Arcachon, France
8-10 December 2008

ISBN: 978-1-61738-299-4

Printed from e-media with permission by:

Curran Associates, Inc.
57 Morehouse Lane
Red Hook, NY 12571

Some format issues inherent in the e-media version may also appear in this print version.

Printed by Curran Associates, Inc. (2010)

For permission requests, please contact the Institute of Physics
at the address below.

Institute of Physics
Dirac House, Temple Back
Bristol BS1 6BE UK

Phone: 44 1 17 929 7481
Fax: 44 1 17 920 0979

techtracking@iop.org

Additional copies of this publication are available from:

Curran Associates, Inc.
57 Morehouse Lane
Red Hook, NY 12571 USA
Phone: 845-758-0400
Fax: 845-758-2634
Email: curran@proceedings.com
Web: www.proceedings.com

TABLE OF CONTENTS

The Anglo-French Physical Acoustics Conference (AFPAC) had its 8th annual meeting at Arcachon in France, between 8 to 10 December 2008. This series of meetings is a successful collaboration between the Physical Acoustics Group (PAG) of the Institute of Physics and the Groupe d'Acoustique Physique, Sous-marine et UltraSonore (GAPSUS) of the Société Française d'Acoustique. First established in 2001, the aim of this annual conference is to provide a forum where the most recent research developments in the field of Physical Acoustics in the UK and France are reviewed. AFPAC alternates between venues in the UK and France and the format has been designed to be 'small and friendly'. The conference attracts the main research leaders in the two countries and likes to encourage research students in particular to have their debut presentations at this event. Every year the organisers also invite a number of keynote speakers who are leading international authorities in their fields. For the meeting in Arcachon, the invited speakers included Dame Anne Dowling of Cambridge University, Anthony Kent of Nottingham University, Michel Bruneau of Université du Maine, Richard Hazelwood, Jérôme Vasseur of Institut d'Electronique, de Micro électronique et de Nanotechnologie and Michel Castaings of Laboratoire de Mécanique Physique of Université Bordeaux 1. The scope of the conference is kept intentionally wide so as to encourage cross-fertilisation of ideas across different application areas in the field of Physical Acoustics. Presentations in the past have covered non-destructive evaluation and material characterisation, medical ultrasound, underwater acoustics and ultrasonic sensors. This is the first time the proceedings of AFPAC are being published in the event's history and the participants were invited to submit manuscripts for publication. As a result there are 9 peer reviewed papers from a total of 34 that were presented at the meeting. In view of the high standard of refereeing that the editors have applied to the received manuscripts, it is hoped that in future years the number of papers submitted for publication in the conference proceedings will grow.

Alain Lhemery
Nader Saffari
Christophe Aristegui

All papers published in this volume of *Journal of Physics: Conference Series* have been peer reviewed through processes administered by the proceedings Editors. Reviews were conducted by expert referees to the professional and scientific standards expected of a proceedings journal published by IOP Publishing.

Journal of Physics: Conference Series **195** (2009) 012001 doi:10.1088/1742-6596/195/1/012001

Simulation of non-destructive inspections and acoustic emission measurements involving guided waves

V Baronian[1,2], A Lhémery[1] and A-S Bonnet-BenDhia[2]

[1] CEA, institut LIST,
centre de Saclay, bât. 611, point courrier 120, 91191 Gif-sur-Yvette *cedex*, France.
[2] POEMS, CNRS-INRIA-ENSTA UMR 7231,
ENSTA ParisTech, 32 boulevard Victor, 75015 Paris, France.

E-mail: alain.lhemery@cea.fr

Abstract. In a structure that guides elastic waves, a discontinuity (defect, shape variation) causes scattering (reflection, partial extinction or mode conversion). Two modal formulations have been developed to link separate models dealing with the calculation of the modal decomposition, with the generation and reception of guided waves (GW), with their scattering. The first concerns pulse-echo configurations (involving a single transducer), the other concerns pitch-catch configurations (two transducers involved). A new finite element (FE) method has been developed to compute the scattering by an arbitrary discontinuity, based on the modal decomposition of the field. Perfectly transparent boundary conditions (Dirichlet-to-Neuman boundaries) are developed, allowing the FE computation zone to be reduced to a minimum. A specific variational problem including these boundary conditions was obtained and solved using FE tools. By combining the modal formulations, the new FE scheme and tools for GW radiation, propagation and reception based on the Semi-Analytical Finite Element (SAFE) method, a new simulation tool has been developed. It can address almost arbitrary configurations of GW nondestructive testing. Moreover, a source inside the FE computation zone can be defined so that configurations of testing by acoustic emission can also be simulated. Examples of use of this tool are shown, some dealing with junctions of complex geometry between two guides, other with surface or bulk sources of acoustic emission.

1. Introduction

The simulation of a complete inspection (a configuration of nondestructive testing) by elastic guided waves requires specific models to describe *i)* the radiation and reception of guided waves by a transducer, *ii)* their propagation and *iii)* their scattering by a flaw or by a variation of the guide shape (a *guide discontinuity* subsequently). Two theoretical formulations have been proposed [1] allowing these phenomena to be computed separately. They result in two overall formulae linking independent models of the various phenomena; the first (resp. second) concerns pulse-echo (resp. pitch-catch) configurations. These formulations are based on a modal description of wave phenomena in the guide. In [1], the method has been derived to deal with waveguides of arbitrary cross section, but the model for computing scattering could only take into account guide discontinuities in the form of planar defects normal to the guide axis.

The object of the present work is to show how the method has been extended to arbitrary discontinuities of the guide. The difficulty is therefore concentrated on the computation of the scattering matrix (such a matrix appears in the two formulae) that links incident modes to either transmitted or reflected modes. An original finite element formulation (FE) has been proposed in [2]; transparent boundary conditions based on modal expansions were derived and used on each side of the discontinuity. This leads to computation zones which can be of very small extent, all the smaller since the number of evanescent modes taken into account is large; this constitutes the main advantage of the method in terms of performances.

The present paper aims at presenting how the various models and formulations can be used together to predict results of nondestructive examination using guided waves. It is organized as follows. First, the various theoretical tools making it possible to predict such results are reviewed: some general results concerning guided modes are briefly recalled, the main point being the biorthogonality relation extensively used subsequently, then the overall modal formulations are described; the FE formulation for computing the scattering of guided waves by an arbitrary scatterer is also presented. Then, examples of results for various configurations are given for demonstrating the versatility of the tools developed. Since source terms can also be included inside or at the surface of the FE computation zone as second members of the system of equations to be solved, the same tools can also be used to model Acoustic Emission testing configurations where transducers are passive; it implies to also model acoustic emission by a defect under stress, which is done here in a very simplified manner. The question of transducer diffraction effects is not addressed here. It is solved in a number of publications for transducers acting at the guiding surface; for transducers acting from the guide section, an original account of typical effects and of the modeling approach to compute them may be found in [3].

2. Theory

2.1. Some results concerning guided waves of specific interest

We consider an isotropic elastic waveguide of section S in the $\mathbf{x}_S = (x_1, x_2)$ plane and of axis x_3 with a stress-free boundary. The density ρ and stiffness coefficients may depend on $\mathbf{x}_S$. The propagation in Ω is modelled by the following classical equations (where $\mathbf{u} = (u_1, u_2, u_3)^T$ denotes the displacement field and $\omega > 0$ the angular frequency):

$$\begin{cases} -\operatorname{div}\boldsymbol{\sigma}(\mathbf{u}) - \omega^2 \rho \mathbf{u} = 0 \quad \text{in}\,\Omega, \\ \boldsymbol{\sigma}(\mathbf{u})\cdot\mathbf{v} = 0 \qquad \text{on}\,\partial\Omega, \end{cases} \tag{1}$$

where $\boldsymbol{\sigma}(\mathbf{u})$, the stress tensor, is related to the strain tensor $\boldsymbol{\varepsilon}(\mathbf{u}) = (\nabla\mathbf{u} + \nabla^T\mathbf{u})/2$ by Hooke's law and $\mathbf{v}$ denotes the outward unitary normal to $\partial\Omega$. It is assumed in equation (1) that there is no source terms in Ω or on $\partial\Omega$. Because the guide may be described as an arbitrary section S translated along an axis (x_3 axis), new notations are more suitable: we denote by $\mathbf{u}_S$, $\mathbf{t}_S$, $\boldsymbol{\sigma}_S$ and $\boldsymbol{\varepsilon}_S$ the transverse part of the displacement field $\mathbf{u}$, the normal stress $\boldsymbol{\sigma}(\mathbf{u})\cdot\mathbf{e}_3$, the stress tensor and the strain tensor, respectively:

$$\mathbf{u}_S = \begin{pmatrix} u_1 \\ u_2 \end{pmatrix}, \mathbf{t}_S = \begin{pmatrix} \sigma_{31} \\ \sigma_{32} \end{pmatrix}, \boldsymbol{\sigma}_S = \begin{pmatrix} \sigma_{11} & \sigma_{12} \\ \sigma_{21} & \sigma_{22} \end{pmatrix}, \boldsymbol{\varepsilon}_S = \begin{pmatrix} \varepsilon_{11} & \varepsilon_{12} \\ \varepsilon_{21} & \varepsilon_{22} \end{pmatrix}, t_3 = -\sigma_{33}\,. \tag{2}$$

Introducing now the following vectors

$$\mathbf{X} = \begin{pmatrix} \mathbf{t}_S \\ u_3 \end{pmatrix}, \mathbf{Y} = \begin{pmatrix} \mathbf{u}_S \\ t_3 \end{pmatrix}, \tag{3}$$

it is possible to re-write the elastodynamic system as an evolution problem (with respect to the coordinate x_3 of the waveguide) expressed for the new variables $\mathbf{X}$ and $\mathbf{Y}$ (by doing so, we extend the formulation proposed in [4, 5] to the general 3D case). On gets

$$\frac{\partial}{\partial x_3}\begin{pmatrix}\mathbf{X}\\ \mathbf{Y}\end{pmatrix}=\begin{pmatrix}0 & F\\ G & 0\end{pmatrix}\begin{pmatrix}\mathbf{X}\\ \mathbf{Y}\end{pmatrix}, \tag{4}$$

where F and G are defined as follows, assuming subsequently that the elastic propagation medium is isotropic (λ and μ denote the Lamé's coefficients). The two operators F and G write

$$F\mathbf{Y}=\begin{pmatrix}-\mathrm{div}_S\,\boldsymbol{\sigma}_\mathrm{S}(\mathbf{Y})-\omega^2\rho\mathbf{u}_\mathrm{S}\\ -\alpha\,\mathrm{div}_S\,\mathbf{u}_\mathrm{S}-\alpha t_3/\lambda\end{pmatrix},\; G\mathbf{X}=\begin{pmatrix}\mathbf{t}_\mathrm{S}/\mu-\nabla_S u_3\\ \mathrm{div}_S\,\mathbf{t}_\mathrm{S}+\omega^2\rho u_3\end{pmatrix}, \tag{5}$$

with $\boldsymbol{\sigma}_\mathrm{S}(\mathbf{Y})=(\delta\,\mathrm{div}_S\,\mathbf{u}_\mathrm{S}-\alpha t_3)\mathbf{I}+2\mu\boldsymbol{\varepsilon}_\mathrm{S}(\mathbf{u}_\mathrm{S})$, and where $\delta=2\lambda\mu/(\lambda+2\mu)$ and $\alpha=\lambda/(\lambda+2\mu)$. Since the vector normal to $\partial\Omega$, $\mathbf{v}=(\mathbf{n}_\mathrm{s}, 0)$, is independent of the axial (x_3) coordinate, the boundary condition can be written equivalently in terms of two uncoupled boundary conditions in **X** and **Y**:

$$\boldsymbol{\sigma}(\mathbf{u})\cdot\mathbf{v}=0 \text{ on } \partial\Omega \Leftrightarrow \begin{cases}\boldsymbol{\sigma}_\mathrm{S}(\mathbf{Y})\cdot\mathbf{n}_\mathrm{S}=0\\ \mathbf{t}_\mathrm{S}\cdot\mathbf{n}_\mathrm{S}=X_1n_1+X_2n_2=0\end{cases} \text{ on } \partial S. \tag{6}$$

Now, a solution of equation (4) can be sought in the form

$$\begin{pmatrix}\mathbf{X}(\mathbf{x}_\mathrm{S})\\ \mathbf{Y}(\mathbf{x}_\mathrm{S})\end{pmatrix}e^{i\beta x_3},\ \beta\in\mathbb{C}, \tag{7}$$

assuming x_3 separable of others coordinates $\mathbf{x}_\mathrm{S}$. This leads to the eigen-value problem that writes

$$i\beta\begin{pmatrix}\mathbf{X}\\ \mathbf{Y}\end{pmatrix}=\begin{pmatrix}0 & F\\ G & 0\end{pmatrix}\begin{pmatrix}\mathbf{X}\\ \mathbf{Y}\end{pmatrix}. \tag{8}$$

A so-called biorthogonality relation can be derived between two solutions $(\beta,\mathbf{X},\mathbf{Y})$ and $(\tilde{\beta},\tilde{\mathbf{X}},\tilde{\mathbf{Y}})$, as demonstrated in due details in [2]

$$\left(FG\mathbf{X}\middle|\tilde{\mathbf{Y}}\right)_S=\left(\mathbf{X}\middle|GF\tilde{\mathbf{Y}}\right)_S\Leftrightarrow(\tilde{\beta}^2-\beta^2)\left(\mathbf{X}\middle|\tilde{\mathbf{Y}}\right)_S=0. \tag{9}$$

In equation (9), the notation $(\bullet|\bullet)$ stands for

$$\left(\mathbf{X}\middle|\mathbf{Y}\right)_S=\int_S(X_1Y_1+X_2Y_2+X_3Y_3)\,dS=\int_S(\mathbf{t}_\mathrm{S}\cdot\mathbf{u}_\mathrm{S}+t_3u_3)\,dS. \tag{10}$$

Equation (9) implies that the eigen-solution $(\beta,\mathbf{X},\mathbf{Y})$ is "orthogonal" to any other mode $(\tilde{\beta},\tilde{\mathbf{X}},\tilde{\mathbf{Y}})$ except if $\tilde{\beta}=\pm\beta$. Let $(\beta_n,\mathbf{X}_n,\mathbf{Y}_n)$ for $n\in\mathbb{N}$ denote the eigen-elements corresponding to right going modes, thus, $(\beta_{-n},\mathbf{X}_{-n},\mathbf{Y}_{-n})=(-\beta_n,-\mathbf{X}_n,\mathbf{Y}_n)$ for $n\in\mathbb{N}$ being the family of left going modes, according to Auld's labeling convention [6]. One gets after suitable normalization:

$$\left(\mathbf{X}_n\middle|\mathbf{Y}_m\right)_S=\delta_{nm}. \tag{11}$$

2.2. Modal formulations for GW nondestructive testing configurations

As it has been mentioned in the introduction of the present paper, two overall formulations to model NDT experiments using guided waves have been derived in [1] for the cases of a single transducer working in the transmit-receive mode and that of two separate transducers for the radiation and the reception. In these models, it is assumed that a waveguide is locally nonuniform, due to the presence of a guide discontinuity. The guide discontinuity is a finite volume Ω_d of the guide bounded by two surfaces Σ^- and Σ^+ as shown by Figure 1. The first configuration results in an expression that is a function of reflection coefficients while the second is expressed as a function of transmission coefficients. Respective expressions for the amplitude received write

$$\delta\Gamma_1 = \frac{-i\omega}{P}\sum_{n\in\mathbb{N}}\sum_{m\in\mathbb{N}} A_n^r A_m^e R_{nm} e^{i(\beta_n+\beta_m)x_3^-}, \quad \delta\Gamma_2 = \frac{-i\omega}{P}\sum_{n\in\mathbb{N}}\sum_{m\in\mathbb{N}} A_n^r A_m^e T_{nm} e^{i\beta_m x_3^-} e^{i\beta_n(L-x_3^+)}, \tag{12}$$

where L denotes the position of the receiver along the x_3-axis, the emitter being located at $x_3 = 0$. $x_3^\pm$ are the coordinates of the two boundaries $\Sigma^\pm$. A_n^e and A_n^r are the amplitude of the various modes emitted (the sensitivity to the modes received) by the transmitting (receiving) transducer. T_{nm} and R_{nm} coefficients denote respectively the corresponding coefficients of reflection from the guide discontinuity and transmission through it, that is, the amplitude of the nth mode reflected or transmitted for the mth incident mode. In (12), it is assumed that the right waveguide in which waves are transmitted is the same as the left one in which the incident waves were propagating. Considering two different guides on each side of Ω_d is however straightforward.

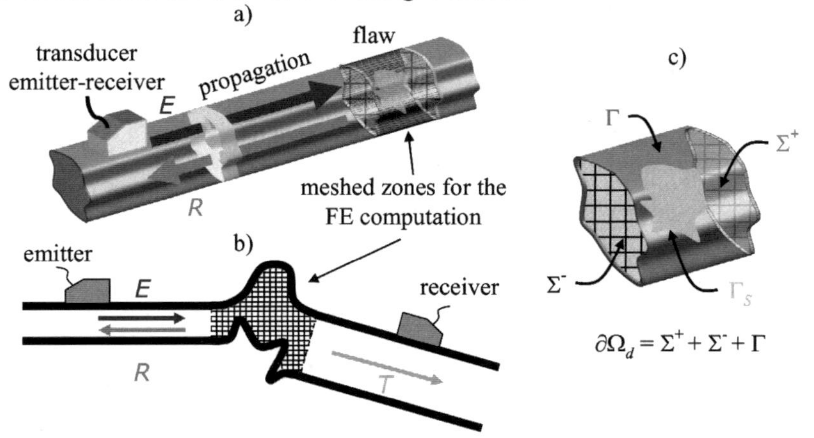

Figure 1. a) and b) represent typical NDT configurations that can be addressed by the overall configurations. c) shows the FE computation zone of a) which is meshed in the volume of the guide limited by the two sections Σ^+ and Σ^-.

In what follows, a FE method is obtained to compute these coefficients, using the various formulae recalled in section 2.1.

2.3. A modal FE scheme for computing the scattering coefficients

The aim here is to derive a FE method based on the modal decomposition of elastic waves in the guide so that the scattering coefficients appearing in equations (12) are readily computed. The interest of doing so is that a finite element method can handle an arbitrary local configuration of scattering in Ω_d between the two boundaries Σ^- and Σ^+ at which the modal solutions for the uniform waveguide are properly defined. This implies that a solution is sought to define boundary conditions on Σ^- and Σ^+ such that these surfaces are transparent for waves diffracted in Ω_d.

For a right-going mode incident on Σ^- given by at $\mathbf{x} = (\mathbf{x}_S, x_3)$

$$\mathbf{u}^{\mathbf{inc}}(\mathbf{x}) = \mathbf{u}^n(\mathbf{x}_S)e^{i\beta_n(x_3-x_3^-)}, \tag{13}$$

the diffracted field $\mathbf{u}^{\mathbf{dif}}(\mathbf{x})$ [with $\mathbf{u}^{\mathbf{dif}}(\mathbf{x}) = \mathbf{u}^{\mathbf{tot}}(\mathbf{x}) - \mathbf{u}^{\mathbf{inc}}(\mathbf{x})$] is sought in the form of a sum of outgoing waves using the modal decomposition in the uniform waveguide, as

$$\mathbf{u}^{\mathbf{dif}}(\mathbf{x}) = \sum_{n\in\mathbb{N}} A_n^{dif\pm}\mathbf{u}^{\pm n}(\mathbf{x}_S)e^{\pm i\beta_n(x_3-x_3^\pm)}, \tag{14}$$

where $A_n^{dif\pm}$ denotes the amplitude of the nth mode of the diffracted field on $\Sigma^\pm$. Then introducing the mixed variables $\mathbf{X}$ and $\mathbf{Y}$ for the diffracted field, these variables can be similarly expanded thanks to the modal decomposition. One gets:

$$\begin{pmatrix} \mathbf{X}^{\mathbf{dif}}(\mathbf{x}) \\ \mathbf{Y}^{\mathbf{dif}}(\mathbf{x}) \end{pmatrix} = \sum_{n\in\mathbb{N}} A_n^{dif\,\pm} \begin{pmatrix} \mathbf{X}^{\pm n}(\mathbf{x}_\mathbf{S}) \\ \mathbf{Y}^{\pm n}(\mathbf{x}_\mathbf{S}) \end{pmatrix} e^{\pm i\beta_n (x_3 - x_3^\pm)}. \tag{15}$$

Using the biorthogonality relation (11), it is possible to extract the amplitude $A_n^{dif\,\pm}$ of the nth mode as

$$A_n^{dif\,\pm} = \left(\mathbf{X}_{\Sigma^\pm}^{\mathbf{dif}} \middle| \mathbf{Y}^{\pm n}\right)_S = \left(\mathbf{Y}_{\Sigma^\pm}^{\mathbf{dif}} \middle| \mathbf{X}^{\pm n}\right)_S. \tag{16}$$

Introducing (16) into (15), one can express two coupling operators (similar to a DtN operator, for Dirichlet-to-Neuman) allowing mapping $\mathbf{X}^{\mathbf{dif}}$ on $\mathbf{Y}^{\mathbf{dif}}$ for the first or *vice-versa* for the second, as:

$$\begin{pmatrix} \mathbf{X}_{\Sigma^\pm}^{\mathbf{dif}} \\ \mathbf{Y}_{\Sigma^\pm}^{\mathbf{dif}} \end{pmatrix} = \sum_{n\in\mathbb{N}} \begin{pmatrix} \left(\mathbf{Y}_{\Sigma^\pm}^{\mathbf{dif}} \middle| \mathbf{X}^{\pm n}\right)_S \mathbf{X}^{\pm n}(\mathbf{x}_\mathbf{S}) \\ \left(\mathbf{X}_{\Sigma^\pm}^{\mathbf{dif}} \middle| \mathbf{Y}^{\pm n}\right)_S \mathbf{Y}^{\pm n}(\mathbf{x}_\mathbf{S}) \end{pmatrix}. \tag{17}$$

Now, using one of these so-called *boundary maps*, one can formulate the diffraction problem in the bounded domain Ω_d, where the boundary conditions on the artificial sections $\Sigma^\pm$ will be expressed by means of the chosen coupling operator. They correspond to an outgoing wave condition for the computational domain Ω_d. This original condition as expressed as a modal formula is of the same nature as what is usually called a transparent boundary condition for bulk elastic waves.

The aim is to propose a variational formulation in displacement. To introduce the coupling operator – equation (17) – **X Y** which are mixed variables in displacement and stress and unknown at present stage, a further and last ingredient must be introduced. This is done in the form of a Lagrange multiplier corresponding to either $\mathbf{t}_\mathbf{S}$ or t_3 depending on the choice made for the boundary map (**X** to **Y** or **Y** to **X**, respectively). Now, **X** or **Y** is known as a function of **u** and of the Lagrange's multiplier. Therefore, the work of external stresses on the artificial boundaries can be expressed within the variational formulation.

Scattering coefficients can now be defined in terms of reflection and transmission coefficients as

$$T_{nm} = \left(\mathbf{Y}^{\mathbf{tot}} \middle| \mathbf{X}_n\right)_{\Sigma^+} / \left(\mathbf{Y}^{\mathbf{inc}} \middle| \mathbf{X}_m\right)_{\Sigma^-} = (A_n^{dif\,+} + \left(\mathbf{Y}^{\mathbf{inc}} \middle| \mathbf{X}_n\right)_{\Sigma^+}) / \left(\mathbf{Y}^{\mathbf{inc}} \middle| \mathbf{X}_m\right)_{\Sigma^-}, \tag{18}$$

$$R_{nm} = \left(\mathbf{Y}^{\mathbf{tot}} - \mathbf{Y}^{\mathbf{inc}} \middle| \mathbf{X}_n\right)_{\Sigma^-} / \left(\mathbf{Y}^{\mathbf{inc}} \middle| \mathbf{X}_m\right)_{\Sigma^-} = A_n^{dif\,-} / \left(\mathbf{Y}^{\mathbf{inc}} \middle| \mathbf{X}_m\right)_{\Sigma^-}, \tag{19}$$

by choosing the **Y** to **X** boundary operator, and where

$$A_n^{dif\,\pm} = \left(\mathbf{t}_\mathbf{S}^{\pm n} \middle| \mathbf{u}_\mathbf{S}^{\mathbf{dif}}\right)_{\Sigma^\pm} + \left(u_3^{\pm n} \middle| t_3^{\mathbf{dif}}\right)_{\Sigma^\pm} = \left(\mathbf{Y}^{\mathbf{dif}} \middle| \mathbf{X}_n\right)_{\Sigma^\pm}. \tag{20}$$

For the tangential stress components, this leads to

$$\mathbf{t}_{\mathbf{S}|\Sigma^\pm}^{\mathbf{tot}} - \delta_{\Sigma\Sigma^-} \mathbf{t}_{\mathbf{S}|\Sigma^-}^{\mathbf{inc}} = \sum_{n\in\mathbb{N}} (A_n^{dif\,\pm} + \delta_{\Sigma\Sigma^+} A_n^{inc}) \mathbf{t}_\mathbf{S}^{\pm n}. \tag{21}$$

with

$$\delta_{\Sigma\Sigma^\pm} = \begin{cases} 0, \text{for } \Sigma = \Sigma^\mp \\ 1, \text{for } \Sigma = \Sigma^\pm \end{cases} \text{ and } A_n^{inc} = \left(\mathbf{Y}^{\mathbf{inc}} \middle| \mathbf{X}_n\right)_{\Sigma^+}. \tag{22}$$

The third component of the displacement, required as a consequence of the choice of the **Y** to **X** boundary operator, is used to complete the formulation as we introduced the t_3 Lagrange multiplier. It similarly writes

$$u_{3|\Sigma^\pm}^{\mathbf{tot}} - \delta_{\Sigma\Sigma^-} u_{3|\Sigma^-}^{inc} = \sum_{n\in\mathbb{N}} (A_n^{dif\,\pm} + \delta_{\Sigma\Sigma^+} A_n^{inc}) u_3^{\pm n}. \tag{23}$$

Equations (21, 23) are used to form the final (mixed) variational formulation where the unknowns are the displacement **u** and the t_3 stress component. Readers interested in full details on the FE formulation are referred to [2]. At this stage, the scattering by an arbitrary guide discontinuity can be predicted for an arbitrary incident field. The numerical scheme can only deal with a finite number of modes. All the propagative modes (a finite set) must be accounted for. The infinite sets of inhomogeneous or evanescent modes are truncated after they have been ordered; amplitude of these modes is a decreasing function of the distance to the scatterer that generates them. Thus, the larger their number accounted for in the formulation, the smaller the computation zone surrounding the scatterer or guide discontinuity. This has been numerically studied and shown in [7]. In our implementation, GW are computed using the Semi-Analytical Finite Element (SAFE) method, a general method based on the early work of Dong and Nelson [8] (infinite layer method as called by these authors) generalised by many authors and still under development to deal with complex cases.

3. Results for GW-NDT configurations

3.1. Introduction

In this first section giving and discussing numerical results, two configurations of NDT by guided waves are considered. The first is an example of application of industrial interest: it concerns the reflection and transmission of guided modes from and through the bonding of two identical plates. The second example deals with a part made of two plates of dissimilar thickness linked by a part of linearly varying section in between. This does not correspond to a specific industrial application; but this exemplifies simulation capabilities of our theoretical model and associated computation code to deal with complex configurations. The propagation of guided waves in several guides of different nature (geometry, material etc.) that are linked together can be predicted. In both cases, the parts considered are supposed to be sound in a first computation; in a second computation, the presence of a flaw (debonding, crack) is assumed. In these results, transducer diffraction effects (either in radiation or in reception) are not taken into account. Thanks to the overall formulations proposed in [1], such effects are easily accounted for by simple but specific amplitude coefficients A^e and A^r.

3.2. Scattering by a bonded region

Figure 2 displays the first configuration considered. Two semi-infinite 2-mm-thick plates made of Aluminium (with bulk wave velocities c_L = 6180 m.s^{-1}, c_T = 3040 m.s^{-1} and density ρ = 2700 kg.m^{-3}) are bonded, either perfectly (first computation) or partially debonded (second).

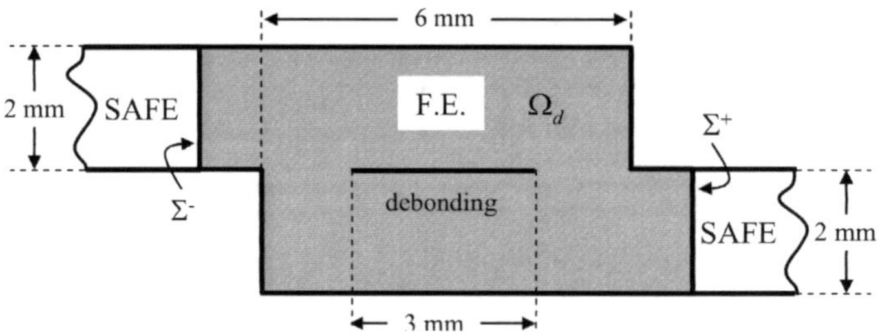

Figure 2. Configurations: two 2-mm-thick plates are bonded (6-mm-overlap). This junction is undamaged in a first computation, partly debonded in a second. The grey region is meshed for the FE computation.

The bonded region is treated as a guide discontinuity, meshed by 27512 triangular elements for the computation of the local FE solution. The same modal solution computed by SAFE (though the classical Lamb wave solution could have been also efficiently computed) is used for both semi-infinite plates. In the frequency bandwidth 0.825 – 1.15 MHz considered, only three modes are propagative. Figure 3 shows the total field in the bond at the frequency of 0.825 MHz, assuming an A0 incident mode in the upper plate. Two components of the particle displacement are shown (2D computations in this example). The transparent boundary operator is applied on $\Sigma^{\pm}$ by taking into account 41 modes.

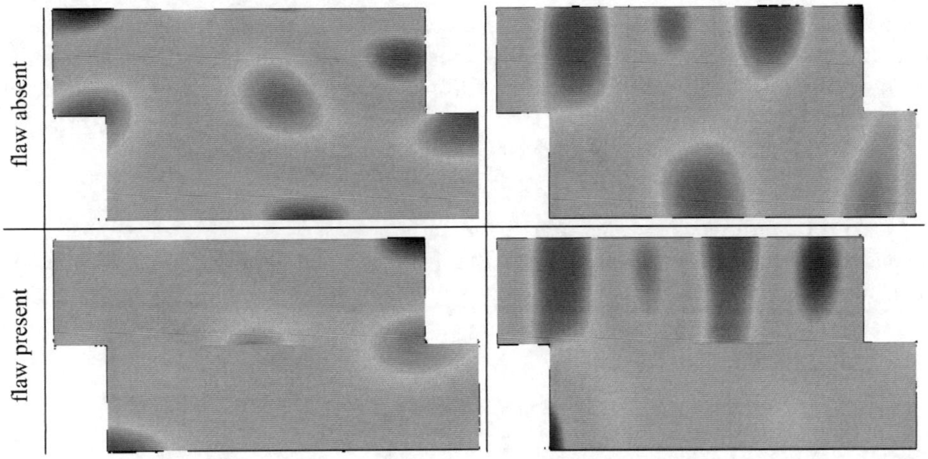

Figure 3. Total fields of particle displacement (x- and y- components in left and right columns, respectively) in the sound bonded region (top) and in the debonded region (bottom); an A0 incident mode is considered, at an excitation frequency of 0.825 MHz.

In these results, the various components in the two cases are not displayed at the same color scale; they should not be quantitatively compared (of course, quantitative results were obtained in the computations). The only interest in presenting them qualitatively here (with a color map maximizing the range between the smallest negative value and the largest positive value) is to show how the incident A0 mode is strongly scattered in the presence of the flaw. Full quantitative results are shown by figure 4 where reflection and transmission coefficients (amplitude of reflected and transmitted modes relatively to that of the incident) are given as functions of frequency in the absence or in the presence of the flaw, again assuming an A0 incident mode in the left plate.

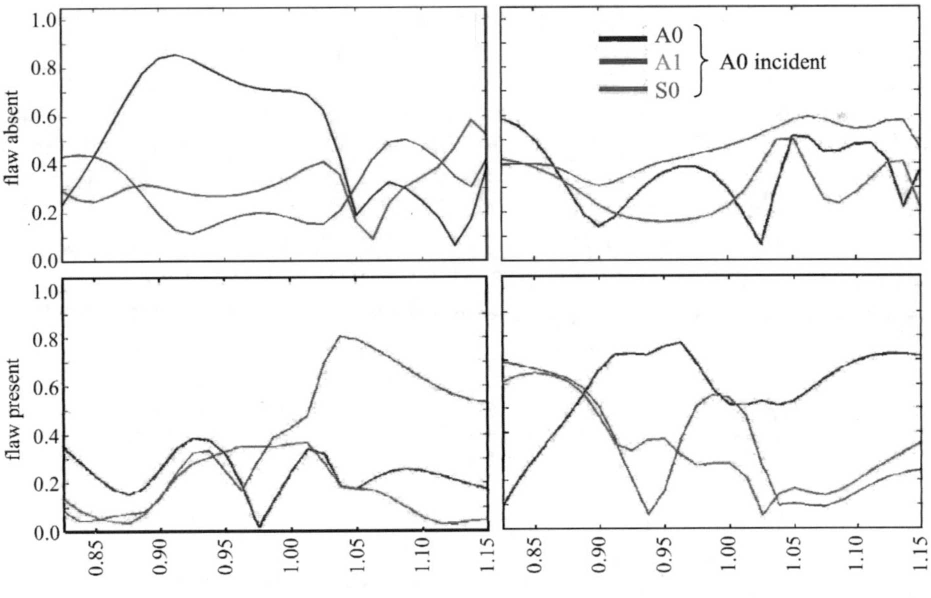

Figure 4. Reflection from (left) and transmission through (right) the bonded zone of the various modes as functions of frequency (in MHz), relatively to an A0 mode incident on the bonded zone, in the absence (top) or presence (bottom) of partial debonding.

Note that by summing the square of the coefficients, one easily checks energy conservation relatively to that carried by the incident mode. Resonances and anti-resonances appear, depending on the mode considered. Such results demonstrate that the prediction of coefficient behaviours cannot be intuitively sought but requires a proper numerical scheme, considering how their variation is complex in this limited bandwidth. Such results together with proper parametric studies (influence of the size, position and nature of the debonding) can be very helpful to design for a given part, the best configuration to maximise the sensitivity of the NDT method to defects sought.

3.3. Scattering by a variation of guide section

Figure 5 displays the second configuration. Two semi-infinite plates of different thicknesses (10- and 20-mm, respectively) are linked by a part of linearly varying thickness. This time, the constitutive material is steel (bulk wave velocities $c_L = 5960$ m.s^{-1}, $c_T = 3260$ m.s^{-1} and density $\rho = 7930$ kg.m^{-3}).

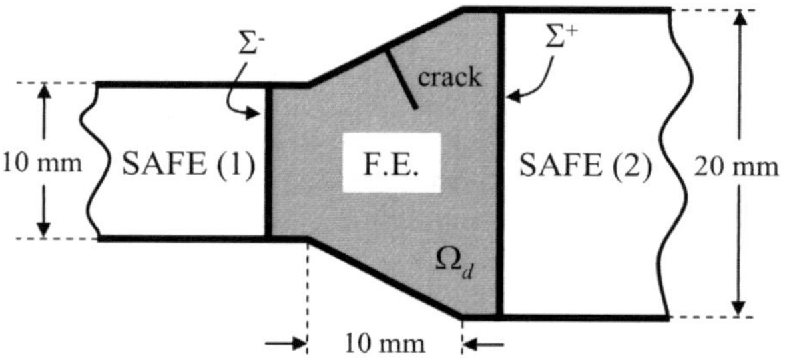

Figure 5. Configurations considered: a thin (10-mm) plate and a thick (20-mm) semi-infinite plate are linked by a part of linearly varying section. In the first computation, this part is sound whereas it is cracked in the second. The grey region is meshed for the FE computation.

The part of varying section is supposed to be undamaged in a first computation and contains a 4-mm-long surface-breaking crack normal to the outer surface in a second one. The intermediate region is again treated as a guide discontinuity; it is meshed by 8688 triangular elements for the computation of the local FE solution using the formulation described in the previous section. The same solution computed by SAFE is used for both semi-infinite plates; since they are of different thickness, the solution is computed over a bandwidth that includes the actual bandwidth of the modelled experience (0.165 – 0.265 MHz) for the incident modes in the thin plate and that (0.33 – 0.53 MHz, frequencies doubled) for dealing with the thick plate which is twice thicker than the first. Corresponding dispersion curves (phase and group velocities) are shown in figure 6.

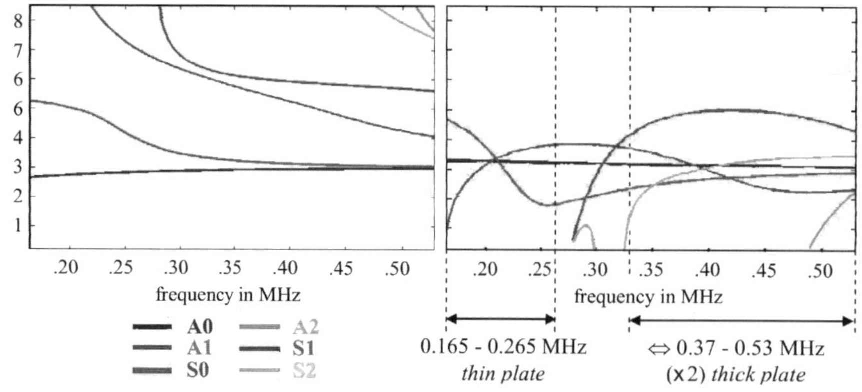

Figure 6. Phase (left) and group (right) velocities [mm.μs^{-1}] in the thin plate of the propagative modes as functions of the frequency for a 10-mm-thick plate.

Figure 7 shows the total field in the part of varying section at the frequency of 0.165 MHz, assuming a S0 incident mode in the thin plate. Two components of the particle displacement are shown (2D computations were also made in this example). In the bandwidth considered, three (six) modes are propagative in the thin (thick) plate. The Dirichlet-to-Neuman transparent boundary operator on $\Sigma^{\pm}$ is applied by taking into account 41 modes. As in the previous case, color amplitude scales are not identical in the various field representations. Due to symmetry breaking in the presence of the crack and the associated abrupt change in cross-sectional area, images show that the total field is strongly affected in such a case as compared to the flawless case.

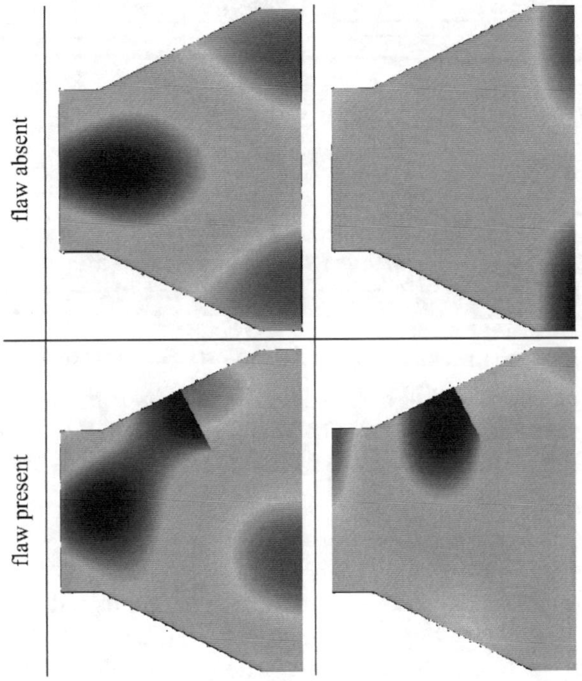

Figure 7. Total fields of particle displacement (x- and y-components in left and right columns, respectively) in the sound region (top) and in the cracked region (bottom); a S0 incident mode is considered, at an excitation frequency of 0.165 MHz.

Full quantitative results are shown by figure 8 where reflection and transmission coefficients (relatively to that of the incident mode) are given in the absence or presence of the flaw. A S0 incident mode in the thin plate is assumed. The various scattering coefficients are given as functions of the frequency, for the three modes A0, A1 and S0 in the thin plate and the six modes (same as in the thin plate plus S1, S2 and A2) in the thick plate.

When the flaw is absent (top results in figure 8), the symmetry of the configuration together with the symmetric incident mode considered leads to null-valued reflection and transmission coefficients for antisymmetric modes, as expected. Again, it is easily checked that the energy conservation criteria are fulfilled by squaring then summing the various coefficients.

In the presence of the surface-breaking crack (bottom results in figure 8), the configuration loses its symmetry: antisymmetric modes are generated by the complex scattering phenomena occurring in the part of varying section that includes the crack. The behaviour is far more complex and requires a proper numerical scheme to be computed.

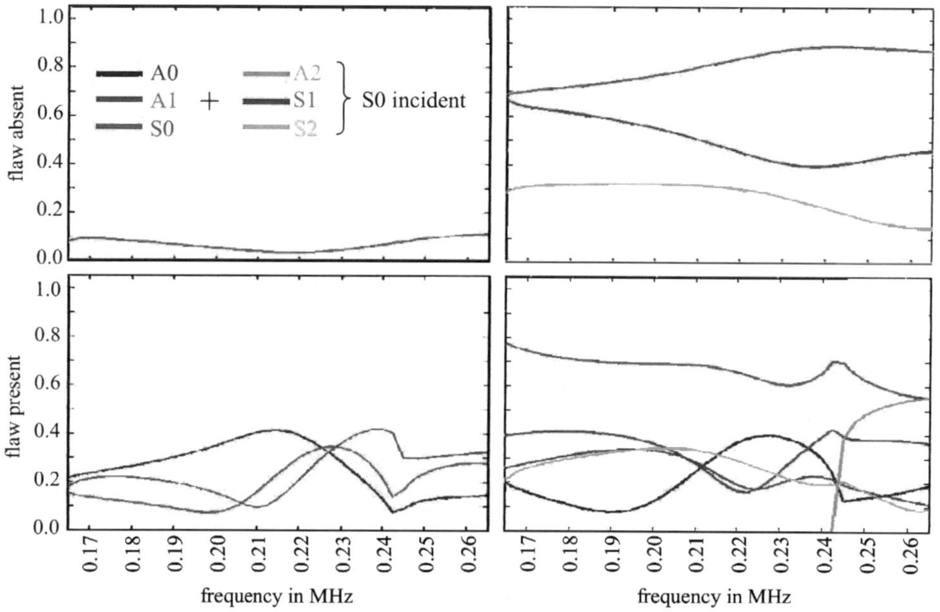

Figure 8. Reflection from (left) and transmission through (right) the zone of section variation of the various propagative modes as functions of the excitation frequency, relatively to a S0 mode incident in the thin plate, in the absence (top) or presence (bottom) of a crack.

4. Early results for NDT by Acoustic Emission

4.1. Introduction

Acoustic Emission (AE) techniques of NDT are used in many industrial domains. Internal defects in pieces submitted to external stress can generate elastic waves. The use of several receivers to catch some of the energy released in the structure as elastic waves allows not only detecting but also locating defects by means of triangulation algorithms. Such algorithms require however the knowledge of the velocity of propagation of the various waves measured. In typical structures for which AE methods are used, waves propagate as guided waves (pressure vessels, pipes etc.), in the frequency bandwidth of the receivers typically used; therefore, the determination of the wavespeed to input in the algorithms for processing a given indication may be a difficult task. In practice, calibration tests are specifically made, one standard source for this being the brittle fracture of a pencil lead at the surface of the piece under test.

Simulation of AE experiments requires the simulation of guided wave propagation in a structure together with that of the source of acoustic emission and that of the reception process by transducers. The simulation tools presented herein, though they were developed more specifically to deal with active NDT using guided waves, can help simulating AE experiments. For this, a necessary step is that consisting in simulating the amplitude of the various wave modes emitted by a given source in the structure. Interestingly, the FE scheme presented in this paper can be modified to do so. The fact is that the FE element code dealing with dynamic elasticity can handle source terms that appear as second members in the system of equations. Basically, the system of equation (1) is now replaced by

$$\begin{cases} -\operatorname{div}\boldsymbol{\sigma}(\mathbf{u}) - \omega^2 \rho \mathbf{u} = f \quad \text{in}\,\Omega, \\ \boldsymbol{\sigma}(\mathbf{u}) \cdot \mathbf{v} = 0 \qquad \text{on}\,\partial\Omega, \end{cases} \tag{24}$$

where f denotes the equivalent force of the acoustic emission source which can concern source points located either inside the piece or at its surface.

Here, we want to demonstrate, in the very simple example of a plate, differences one can observe depending on the nature and location of the source, therefore, to demonstrate the usefulness of a proper simulation tool to predict which wave modes are potentially measured in an experiment. Note that in this example, the bulk AE source is not realistic. However, more realistic sources can easily be introduced in the FE computation; specific work is in progress to model realistic AE sources in the frame of simulation tools presented in this paper.

4.2. Numerical results

Two sources are considered as depicted by figure 9. The first is a set of three point forces acting at or close to the centre of a 20-mm-thick steel plate directed along the propagation axis. The second is a point source of normal force due to a pencil lead break acting on one of the guiding surface.

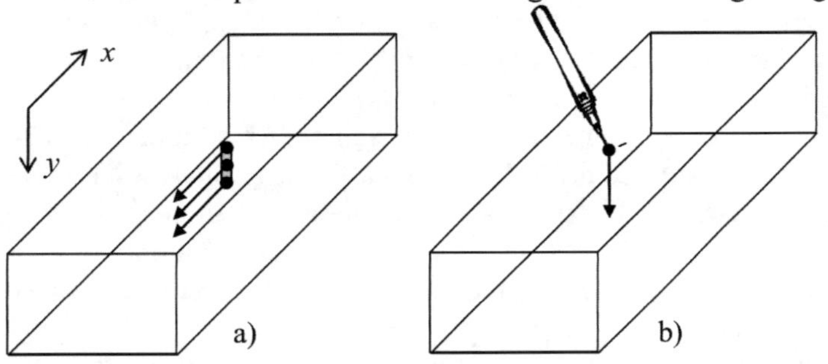

Figure 9. Two sources of acoustic emission are considered: a) an internal source of horizontal force, an external source (lead break).

2D computations have been made. A 2D region of the 20-mm-thick plate (20 x 20 mm²) in which the two sources act has been meshed by 24728 elements, though a smaller zone could have been considered. It is assumed that receivers are sensitive in a narrow bandwidth centred on the frequency 0.818 MHz. This rather high frequency was chosen in this example so that many (17) propagative modes coexist. In the common practice, lower frequencies are used for such a thick plate just to avoid this. Knowledge of higher modes radiation could be used for more accurate source characterization; this is not a conclusion of present work but may constitute an objective for future researches. Dirichlet-to-Neuman transparent boundary conditions were applied on a set of 41 modes. Tables 1 and 2 give the amplitude obtained for the various modes (normalized in each case relatively to the highest amplitude value). Table 1 (2) concerns (anti)symmetric modal decomposition. The behaviour is very different in the two cases. A remarkable fact is that anti-symmetric modes are almost not radiated by the lead break. The relative amplitudes of A0 and S0 modes in the internal source case are also remarkably equal. More importantly, as far as application to acoustic emission testing is concerned, it must be noticed that modes of highest amplitude are different in the two cases. Since they propagate at different wavespeeds, if the second case be used for calibrating the experience, the wavespeed deduced being used in triangulation algorithms, errors could easily occur in source location.

Table 1. Relative amplitudes of modes (S-modes) emitted by the two sources of acoustic emission considered in the plate, for a receiver working at the frequency of 0.818 MHz.

modes	**S0**	**S1**	**S2**	**S3**	**S4**	**S5**	**S6**	**S7**	**S8**
internal	1.00	0.25	0.37	0.48	0.13	0.19	0.41	0.62	0.17
lead break	0.00	0.26	0.44	0.52	0.55	0.52	0.82	0.06	1.00

Table 2. Relative amplitudes of modes (A-modes) emitted by the two sources of acoustic emission considered in the plate, for a receiver working at the frequency of 0.818 MHz.

modes	**A0**	**A1**	**A2**	**A3**	**A4**	**A5**	**A6**	**A7**
internal	1.00	0.14	0.32	0.43	0.48	0.22	0.09	0.40
lead break	0.00	0.00	0.01	0.03	0.04	0.03	0.02	0.05

5. Conclusion

A numerical method has been developed for simulating NDT experiments based on elastic guided waves. This method relies on a modal formulation [1] to deal with the radiation by an ultrasonic transducer, the guided propagation, the scattering by a defect and the reception by the same transducer (pulse-echo configurations) or another one (transmission configurations), each stage or phenomenon being computed separately. The modal formalism is particularly interesting to easily interpret typical complicated waveforms observed in experiments; furthermore, number of interesting results can be obtained by means of simple post-processing, thus limiting computational efforts in multi-parametric studies. In practice, eigenmodes are computed by means of the well-established SAFE method for numerical efficiency and versatility.

A specific Finite Element model has been derived to compute in the smallest possible zone the scattering of an incident field by an arbitrary defect or generally speaking by a discontinuity of the guide; it combines exact transparent boundary conditions and the account of the modal decomposition of an arbitrary field thanks to biorthogonality relations [2]. Further, this FE scheme can admit a second member term allowing one to account for source terms inside the FE computation zone; it is thus possible to simulate also an Acoustic Emission experiment involving guided wave propagation. These developments result in a versatile and numerically efficient method that can address complicated configurations typical of those encountered in practice. At present, the numerical tool developed is used to deal with various configurations of interest for NDT applications.

6. References

[1] Jezzine K and Lhémery A 2007 Simulation of guided wave inspection based on the reciprocity principle and the semi-analytical finite element method *Review of Progress in QNDE* **26**, ed D O Thompson and D E Chimenti (Melville: AIP Conf. Proc. **894**) 39

[2] Baronian V, Bonnet-Ben Dhia A-S and Lunéville E 2009, Transparent boundary conditions for the harmonic diffraction problem in an elastic waveguide, to appear in *J. Comp. Appl. Math.* (preprint already published on-line).

[3] Jezzine K and Lhémery A 2006 Diffraction effects on ultrasonic guided waves radiated or received by transducers mounted on the section of the guide *Review of Progress in QNDE* **25**, ed D O Thompson and D E Chimenti (Melville: AIP Conf. Proc. **820**) 134

[4] Pagneux V and Maurel A 2004 Scattering matrix properties with evanescent modes for waveguides in fluids and solids *J. Acoust. Soc. Am.* **116** 1913

[5] Pagneux V and Maurel A 2006 Lamb wave propagation in elastic waveguides with variable thickness *Proc. R. Soc. A* **462** 1315

[6] Auld B A 1990 *Acoustic fields and waves in solids - Vol. II* (Malabar: Krieger Publishing Company) Chapter 10

[7] Baronian V, Bonnet-Ben Dhia A-S, Lhémery A and Lunéville E 2009 A numerical method for the simulation of NDT experiments in an elastic waveguide *Ultrasonic wave propagation in non homogeneous media*, ed A Léger and M Deschamps (Springer) 137

[8] Dong S B and Nelson R B 1972 On natural vibrations and waves in laminated orthotropic plates *J. Appl. Mech.* **39** 739

Acoustic modelling in view of a determination of the Boltzmann constant within 1 ppm for the redefinition of the kelvin

Pierre Gélat[1], Nicolas Joly[2], Michael de Podesta[1,†], Gavin Sutton[1], Robin Underwood[1]

[1] National Physical Laboratory, Hampton Road, Teddington, TW11 0LW, United Kingdom.

[2] Laboratoire d'Acoustique de l'Université du Maine, UMR CNRS 6613, Avenue O. Messiaen, 72085 Le Mans, cedex 9, France.

[†] E-mail: Michael.dePodesta@npl.co.uk

Abstract. iMERA/Euromet Project 885 is co-ordinating European effort towards a new determination of the Boltzmann constant k_B to within 1 ppm with the aim of redefining the unit of thermodynamic temperature. This project will enable the National Physical Laboratory to perform primary thermometry in the region of –40 °C (Hg) to 156 °C (In) with sub-millikelvin uncertainties by 2012. The chosen technique relies on determining the speed of sound in a monatomic gas. Using the radial acoustic modes of a spherical resonator, consisting of a copper shell and filled with argon or helium, the speed of sound can be measured with great precision and from this measurement the Boltzmann constant can be inferred. This project draws on expertise in dimensional, density, microwave and acoustic measurements at the state-of-the-art. In order to gain further understanding of the experimental configuration a vibro-acoustic model has been developed using the finite element method. Initial calculations were carried out to ensure that predictions of the resonant frequency could be made with the required precision by comparing against an analytical model of a spherical shell filled with a gas. A more elaborate model better representing the experimental configuration was then developed. Thermo-viscous effects close to the fluid-structure boundary were accounted for using a linear acoustic formulation, from which a normal incidence admittance boundary condition was derived and imposed on the inner surface of the resonator. Acoustic pressure, particle velocity and temperature variation as a function of position may be obtained within the gas as a function of frequency. It is therefore possible to investigate how changes in the configuration affect the frequency of radial modes. It is hoped that this approach will shed a better understanding of the underlying complex physical phenomena allowing a minimization of the overall uncertainty.

1. Background

One aim of fundamental metrology is to provide a system of measurement in which the units of measurement are stable in time and space. To this end, it is preferable that the definitions of the units of measurement do not refer to any specific artefact or substance. Instead, they should relate the units to immutable constants of nature. This work was undertaken as part of a project to replace the current definition of the unit of temperature with a definition based on the Boltzmann constant k_B.

The previous most accurate determination of the Boltzmann constant was made by Moldover *et al* [1] at National Institute of Standards and Technology (Gaithersburg, MD, USA) using a spherical acoustic resonator. This yielded an uncertainty in k_B of ±1.8 ppm with a coverage factor of one. After reviewing possible techniques, the National Physical Laboratory (Teddington, UK) has chosen to again use a spherical acoustic resonator technique, but updating the experiment with 30 years of improved technology. In this experiment, the speed of sound is determined by measuring the resonance frequencies of a dimensionally characterised resonator at the triple point of water (0.01 °C). The frequencies are measured at a number of pressures, and the zero pressure speed of sound inferred. The experiment is conducted using isotopicaly and chemically analysed pure argon. From this result k_B is inferred using:

$$k_{\mathbf{B}} = \frac{M c_0^2}{\gamma N_{\mathbf{A}} T} \tag{1}$$

where

- M is the molar mass with isotopic corrections
- c_0 is the zero-pressure speed of sound
- γ is the ratio of ideal gas heat capacities
- N_A is the Avogadro constant
- T is the thermodynamic temperature.

By examining the large number of factors which must be considered in order to achieve the required level of uncertainty (i.e. within 1 ppm), the interaction of the acoustic modes with the vibrational modes of the shell were identified as being of key importance. An analytical theory exists for the interaction of a uniform spherical shell with a resonating fluid [2],[3], but the effects of shell motion are considered to be independent of thermo-viscous boundary layer corrections. Although efforts are made to include effects due to imperfect resonator geometry using perturbation theory, the experimentally observed vibrational spectra of spheres are actually much more complex than predicted by [2],[3].

Joly *et al* [4] developed a formulation of linear acoustics in thermo-viscous fluids at rest in the form of a set of two coupled equations for particle velocity and temperature variation. In the case of normal local incidence, which is closely approximated during a breathing mode in a spherical resonator, an expression for the specific acoustic admittance at the fluid/structure interface may be obtained.

As part of the work described in this paper, in order to further understand the complex thermal, acoustical and vibrational interactions which occur when a gas-filled spherical resonator is subjected to an acoustic excitation, the acoustic admittance boundary condition described in [4] was used in an acoustic boundary element formulation coupled to the inner wall of the resonator shell. The latter was modelled using the finite element method in order to allow more complex geometries to be examined. Once the vibro-acoustic finite element analysis is completed with the thermo-viscous boundary condition in place, shell displacements can then serve as input quantities to the coupled set of equations described in [4] and the temperature and particle velocity variations within the resonator may be obtained.

This paper outlines the initial stages of this analysis of spherical gas-filled resonators. It is understood that this work will be expanded upon and will form the basis for a more in-depth paper.

2. Experimental configuration

Practically it is impossible to create a perfect spherical resonator and real resonators used experimentally are far from spherical. It is the need to extend our understanding to real resonators which motivates this work. For the radial acoustic modes (which are the main focus of our work) the perturbations from ideal spherical resonator values are affected by two types of shell interaction: *accidental* interactions with complex shell modes which cause anomalous readings at one particular frequency (corresponding to one particular temperature); and interactions with the *breathing mode* of the shell, which will systematically affect all data, even data taken at frequencies far from the breathing mode. It is this systematic effect that we are seeking to evaluate.

Externally resonators differ from perfect spherical shells in three ways. Firstly they must be held in some way, typically by a neck which additionally provides a thermally conducting path to a temperature bath. Secondly the resonators are manufactured from two hemispheres and then (usually) bolted together resulting in a potential change in elastic parameters at the equatorial plane. Finally the thickness of the shell is not uniform, there being typically some kind of cylindrical flange around the equatorial seam. All of these deviations from sphericity mean that a pure radial acoustic mode within the cavity gives rise to non-uniform shell modes, and so couples not just to the breathing mode, but also to other (generally lower frequency) shell modes.

In the experiments we are seeking to model, the sphere is held tightly by its neck in contact with a long steel post holder and secured inside a cylindrical stainless steel pressure vessel. The entire vessel is immersed inside a bath of anti-freeze at approximately the triple point of water. The arrangement is sufficient to keep the temperature within 10 mK of the triple of the point of water, with drift rates less than 0.1 mK per minute. The configuration is shown in Figure 1 (a) and details of the resonator dimensions are shown in Figure 1 (b).

3. Method

In order to accurately compute the coupled vibration of the shell and acoustic field inside the spherical cavity filled with a thermo-viscous fluid, the system is modelled in two steps. In a first vibroacoustic model (Section 3.1), the vibration of the shell is computed whilst accounting for the acoustic coupling, by using a potential pressure formulation for acoustics and where thermo-viscous boundary effects are accounted for by an admittance-like condition. In a second model (Section 3.2), the full solution for thermo-viscous acoustics is computed in the fluid domain, according to the vibrating motion of the shell boundaries of the cavity.

3.1. Vibroacoustic modelling

The vibroacoustic modelling described in this section was carried out using the PAFEC (Program for Automatic Finite Element Calculations) finite element software. The shell was modelled using axisymmetric isoparametric quadratic elements. The gas inside the resonator was modelled using an interior boundary element formulation. The fluid inside the cavity was excited using a point source placed on the axis of symmetry 3.24 cm from the centre of the sphere on the neck side of the resonator. It should be noted that this position was arbitrarily chosen pending a more realistic description of the acoustic source. Harmonic excitation is assumed. The solution for the acoustical analysis was obtained using a surface Helmholtz formulation using the Burton-Miller method [5]. A specific acoustic admittance condition is imposed on the boundary element which takes into account the thermo-viscous boundary layer effects. This expression is described in further detail in [4]. The shell properties are those of copper and the gas properties are those of Argon. The fluid loading at the resonator exterior wall has been neglected in this analysis, as initial modelling has suggested that accounting for this does not affect the pressures inside the cavity by an appreciable amount in the context of the uncertainties required here.

Figure 2 shows the finite element mesh generated to carry out the finite element analysis. The resonator also comprises an equatorial belt as well as a "neck", which are straightforward to include into the analysis, and which will cause the shell to displace in ways that differs from a uniform thickness resonator. At this stage, no mechanical restraints have been implemented.

The properties of argon at the triple point of water are displayed in table 1. These have been obtained from [6].

Table 1. Properties of argon at 0.01°C, 1 atm

Property	**Value**
Equilibrium density ρ_0	1.783890 kg m^{-3}
Speed of sound c_0	307.8623 m s^{-1}
Shear viscosity μ	2.0984×10^{-5} Pa s
Bulk viscosity η	0
Thermal conductivity λ	0.01641 W m^{-1} K^{-1}
Specific heat capacity at constant pressure C_p	521.845 J kg^{-1} K^{-1}

The properties of the copper wall are displayed in table 2.

Table 2. Properties of copper at 0.01°C

Property	**Value**
Density ρ_w	8933 kg m^{-3} [7]
Poisson's ratio ν_w	0.343 [8]
Young's modulus E_w	129.8 GPa [8]
Thermal conductivity λ_w	402.93 W m^{-1} K^{-1} [9]
Heat capacity per unit mass C_w	379.20 J kg^{-1} K^{-1} [9]

From the solution described in [2],[3], it can be found that, for a constant shell thickness gas-filled spherical resonator with the properties outlined in figure 1 and in tables 1 and 2, a radial mode exists around 4380 Hz. By sweeping the frequency of excitation around an interval centred on 4380 Hz, a local maximum for the spatial average of the RMS acoustic pressure at the resonator wall can be found, as shown in figure 3. This maximum is found to occur at 4379.523 Hz.

A plot of the maximum structural deformations is shown in figure 4. It should be noted that the values obtained for the acoustic pressures inside the cavity inside the acoustic display element mesh region are approximate. Although the acoustic loading on the resonator wall resulting from thermo-viscous boundary layer effects is accounted for, this section of the analysis only aims to predict the shell displacements and the acoustic pressures at the resonator wall. Nevertheless, this provides some guidance as to how the fluid inside the cavity behaves, as thermo-viscous effects further away from the wall are likely to be negligible.

3.2. Thermo-acoustic modelling

This subsection contains a brief overview of the formulation described in [4]. The fluid is assumed to be Stokesian, which implies that the stress is proportional to the rate of strain and the heat flux is proportional to the temperature gradient. The fluid is assumed to be homogeneous and at rest, with a linear behaviour for the differential form of its equation of state. When accounting for the phenomenological complementary relationships of the fluid, the fundamental laws of mechanics take the following linearised form.

(i) Linearised Navier-Stokes equation:

$$\rho_0 \frac{\partial \vec{u}}{\partial t} + \vec{\nabla} p - \rho_0 c_0 l_v \vec{\nabla}(\nabla \cdot \vec{u}) + \rho_0 c_0 l_v^{'} \vec{\nabla} \times \vec{\nabla} \times \vec{u} = \vec{0} \tag{2}$$

where the viscous characteristic lengths are given by:

$$l_v = \frac{1}{\rho_0 c_0}\left(\eta + \frac{4}{3}\mu\right) \tag{3}$$

$$l_v^{'} = \frac{\mu}{\rho_0 c_0} \tag{4}$$

(ii) Conservation of mass:

$$\frac{\partial \rho^{'}}{\partial t} + \rho_0 \nabla \cdot \vec{u} = 0 \tag{5}$$

(iii) Conservation of energy:

$$\rho_0 T_0 \frac{\partial s}{\partial t} - \rho_0 c_0 C_p l_h \nabla \cdot \left(\vec{\nabla} \tau\right) = 0 \tag{6}$$

where the thermal characteristic length is given by:

$$l_h = \frac{\lambda}{\rho_0 c_0 C_p} \tag{7}$$

The symbols have the following meaning.

- ρ' is the density variation
- p is the acoustic pressure
- $\vec{u}$ is the particle velocity vector
- s is the entropy variation per unit mass
- τ is the temperature variation
- T_0 is the ambient temperature

The thermodynamic state laws of the fluid can be used to express thermodynamic quantities with two independent variables: p and τ. The equations of state of the gas then become

$$\rho' = \frac{\gamma}{c_0^2}\left(p - \hat{\beta}\tau\right) \tag{8}$$

$$s = \frac{C_p}{T_0}\left(\tau - \frac{\gamma - 1}{\hat{\beta}\gamma} p\right) \tag{9}$$

where

$$\hat{\beta} = \left(\frac{\partial P}{\partial T}\right)_V = \sqrt{\frac{(\gamma - 1)\rho_0^2 c_0^2 C_p}{\gamma^2 T_0}} \tag{10}$$

and γ is the ratio of specific heats.

By combining (2), (5), (6), (8) and (9) and eliminating p, s, and ρ' and expressing the result in terms of $\vec{u}$ and τ, a system of two coupled equations involving $\vec{u}$ and τ is obtained. This system can be solved numerically by analogy with elastodynamic and thermal problems.

$$-\frac{\partial^2 \vec{u}}{\partial t^2} + \left(\frac{c_0^2}{\gamma} + c_0 l_v \frac{\partial}{\partial t}\right)\vec{\nabla}(\nabla \cdot \vec{u}) - c_0 l_v' \frac{\partial}{\partial t}\left(\vec{\nabla} \times \vec{\nabla} \times \vec{u}\right) - \frac{\hat{\beta}}{\rho_0}\frac{\partial}{\partial t}\left(\vec{\nabla}\tau\right) = \vec{0} \tag{11}$$

$$\frac{\partial \tau}{\partial t} - \gamma l_h c_0 \nabla \cdot \vec{\nabla}\tau + \frac{\gamma - 1}{\gamma\hat{\beta}}\rho_0 c_0^2 \nabla \cdot \vec{u} = 0 \tag{12}$$

Using the shell axial and radial velocities of the inner wall as input quantities to the thermo-acoustic code at the Laboratoire d'Acoustique de l'Université du Maine (Le Mans, France) based on equations (11) and (12), temperature and particle velocity variations inside the resonator were obtained during the course of one acoustic cycle. Figure 5 shows the axisymmetric temperature variation inside the cavity. Figure 6 shows the particle velocity vector variation.

4. Discussion

A comprehensive model which aims to include vibrational, acoustical and thermal phenomena in spherical gas-shaped resonators, has been developed and one particular configuration has been investigated. Of all the assumptions that have been made that of axisymmetry of the resonator is perhaps the most controversial. The internal surface is in fact not spherical, but carefully manufactured in the shape of a triaxial ellipsoid, a shape with no rotational symmetry about any axis. Additionally the sphere is manufactured as two hemispheres which are held together by steel bolts. This combined with acoustic excitation which is, in practice, potentially non-axisymmetric, implies that many modes which exists in the experimental device will not be predicted by the model here. A full three-dimensional analysis would be required for this, but this is beyond the scope of this work.

The vibroacoustic analysis centred around 4380 Hz shows that, despite having selected a frequency of excitation corresponding to a radial mode, the shell displacements do not exhibit a breathing mode-like behaviour. Nevertheless, the acoustic display elements in figure 4 as well as the temperature and particle velocity variations in figures 5 and 6 show that the fluid is clearly displaying radial mode behaviour inside the cavity. There is therefore clearly modal interaction between the (0,2) radial mode and the (3,1) non-radial mode, as shown in figure 3.

At this stage, a point source of unit source strength along the axis of symmetry has been used to generate the acoustic excitation. This does not detract from predicting features of how the resonator behaves, given that linearity is assumed. This methodology however needs to be refined to include some features of the emitting transducer, perhaps initially in the form of a rigidly vibrating plane piston, so as to ensure that the acoustic pressures and resulting temperature variations are of the order of what can be measured experimentally.

It should be noted that the admittance boundary condition described in [4] was implemented in the vibroacoustic model (Section 3.1) for local normal incidence, because of the shape of breathing modes in spherical resonators. It accurately accounts for the thermo-viscous acoustic boundary layers effects in the first numerical vibroacoustic model.

5. Conclusions

A fully coupled axisymmetric vibrational, acoustical and thermo-viscous model of a spherical resonator was described in this paper. This model represents an improvement on prior ones given that it enables the shell dynamics to be simulated using the finite element method and therefore allows for more complex geometries to be investigated and to determine their impact on thermo-acoustic behaviour inside the cavity. This model is intended to serve as a design and analysis tool to gain further insight into the complex multi-physics of the behaviour of gas-filled resonators used for the measurement of thermophysical properties. It is hoped that it will help assist further measurements and aid the design of an improved resonator.

References

[1] Moldover MR, et al, *Measurement of the Universal Gas Constant R Using a Spherical Resonator.* Physical Review Letters, 1988. **60**(4): p. 249-252.

[2] Mehl JB, *Spherical Acoustic Resonator: Effects of Shell Motion.* Journal of the Acoustical Society of America, 1985. **78**(2): p. 782-788.

[3] Moldover MR, Mehl JB and Greenspan M, *"Gas-filled resonators: Theory and experiment"*, JASA **79**(2), 1986.

[4] Joly N, Bruneau M and Bossart R, "Coupled equations for particle velocity and temperature variation as the fundamental formulation of linear acoustics in thermo-viscous fluids at rest", Acta Acustica **92** 202-209, 2006.

[5] Burton AJ and Miller GF "The applications of integral equation methods to the numerical solution of some exterior boundary-value problems" Procs of the Royal Society of London A323 issue 1553 1971 pp 210-210.

[6] REFPROP - REFerence fluid PROPerties software, NIST. Database 23 V.8. (2007)

[7] Kaye GWC and Laby TH 1995 *Tables of Physical and Chemical Constants*, Longman Group Ltd., Essex, UK, p. 213.

[8] Kaye GWC and Laby TH 1995 *Tables of Physical and Chemical Constants*, Longman Group Ltd., Essex, UK, p. 44.

[9] Kaye GWC and Laby TH 1995 *Tables of Physical and Chemical Constants*, Longman Group Ltd., Essex, UK, p. 82.

Journal of Physics: Conference Series **195** (2009) 012002 doi:10.1088/1742-6596/195/1/012002

(a)

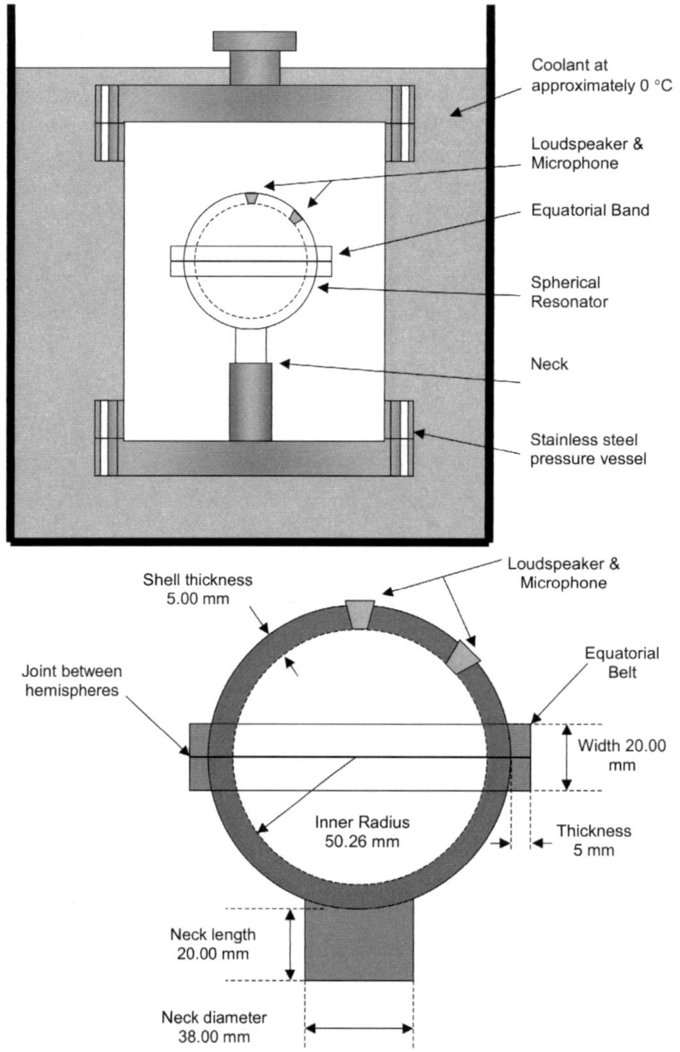

(b)

Figure 1 (a) Schematic arrangement of the resonator within its pressure vessel. (b) Details of the resonators dimensions

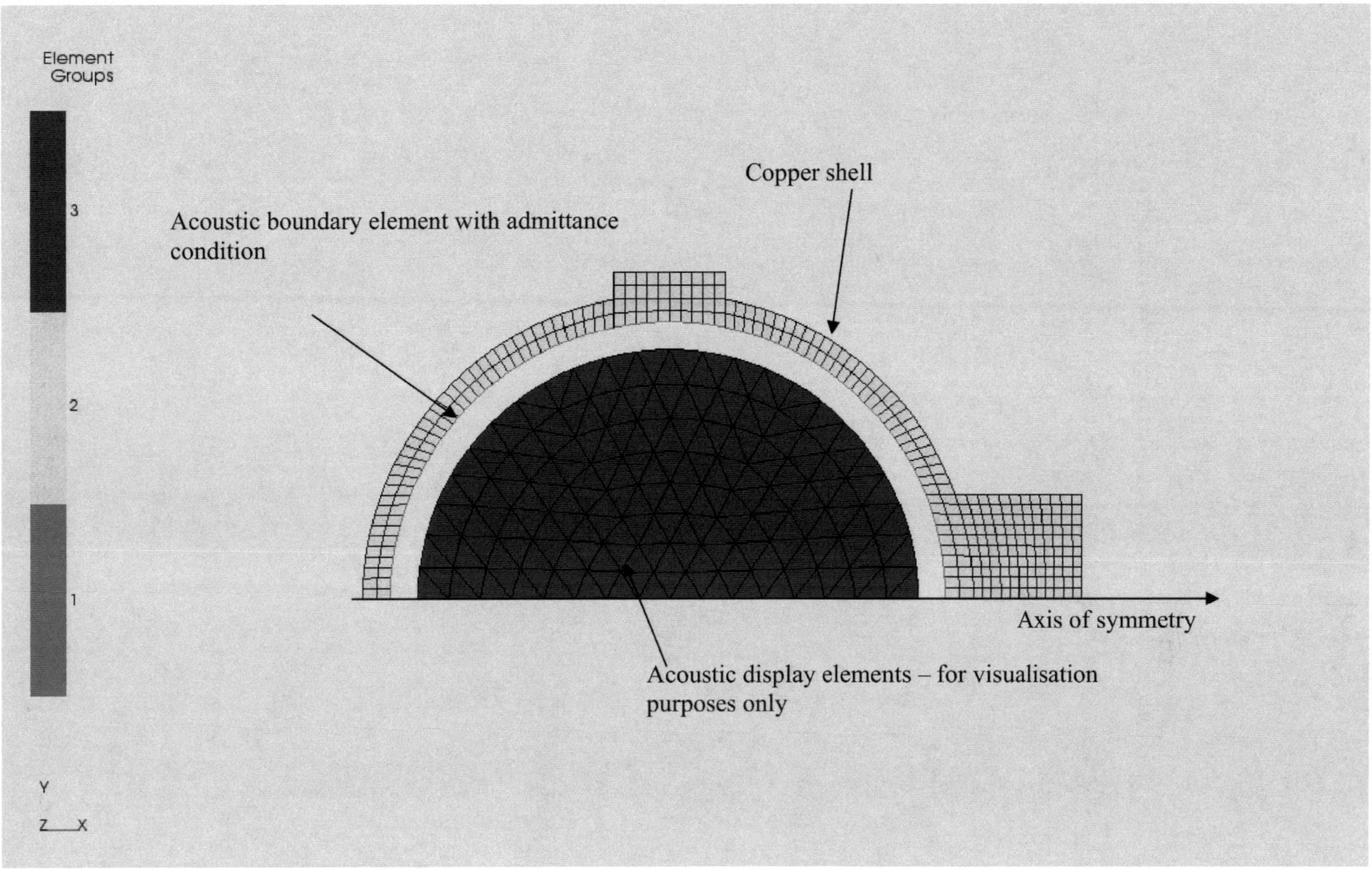

Figure 2 – Finite element mesh used for vibroacoustic analysis.

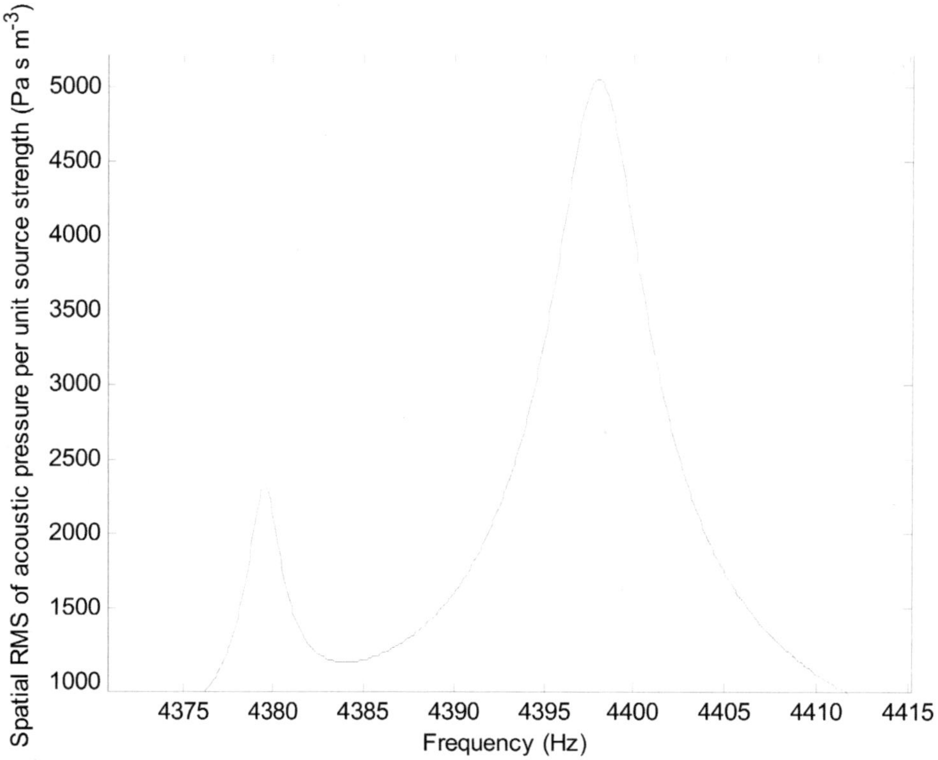

Figure 3 – Spatial RMS value of acoustic pressure per unit source strength at resonator wall as a function of frequency.

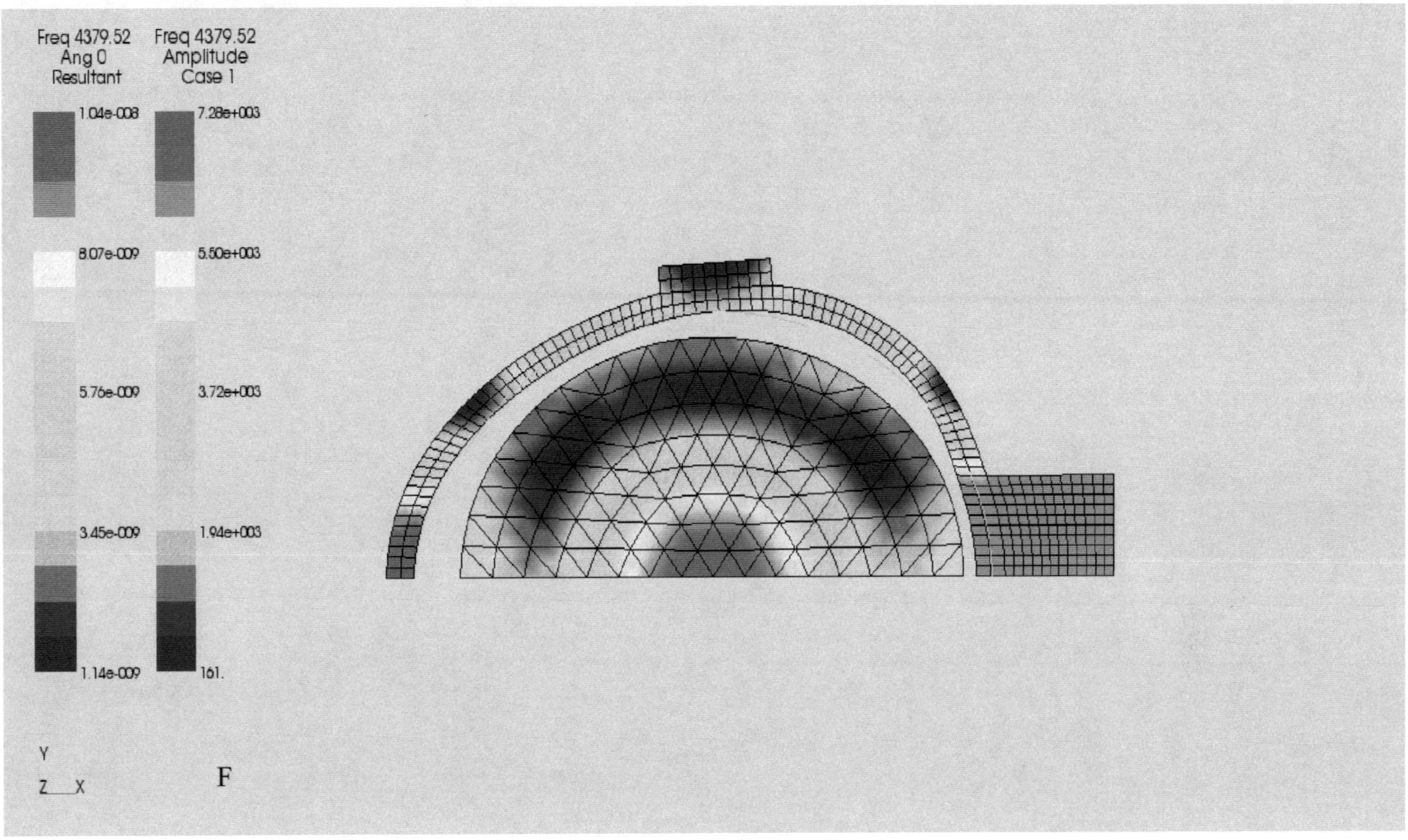

igure 4 – Maximum structural deformations resulting from acoustic excitation at 4379.523 Hz http://www.univ-lemans.fr/%7Enjoly/sphere_Ar/T_variation.gif

Figure 5 – Acoustic perturbation due to the shell motion: temperature variation (colour map: temperature variation)

http://www.univ-lemans.fr/%7Enjoly/sphere_Ar/particle_V.gif

Figure 6 – Acoustic perturbation due to the shell motion: particle velocity (colour map: modulus, arrows: orientation)

Fatigue crack growth monitoring in multi-layered structures using guided ultrasonic waves

E Kostson and P Fromme

Department of Mechanical Engineering, University College London, WC1E 7JE, UK

e-mail: e_kostson@meng.ucl.ac.uk; p.fromme@ucl.ac.uk

Abstract. This contribution investigates the application of low frequency guided ultrasonic waves for monitoring fatigue crack growth at fastener holes in the 2nd layer of multi-layered plate structures, a common problem in aerospace industry. The model multi-layered structure investigated consists of two aluminum plate-strips adhesively bonded using a structural paste adhesive. Guided ultrasonic waves were excited using multiple piezoelectric discs bonded to the surface of the multi-layered structure. The wave propagation in the tensile specimen was measured using a laser interferometer and compared to numerical simulations. Thickness and width mode shapes of the excited flexural waves were identified from Semi-Analytical Finite Element (SAFE) calculations. Experiments and 3D Finite Element (FE) simulations show a change in the scattered field around fastener holes caused by a defect in the 2nd layer. The amplitude of the guided ultrasonic wave was monitored during fatigue experiments at a single point. The measured changes in the amplitude of the ultrasonic signal due to fatigue crack growth agree well with FE simulations.

Keywords: Guided Ultrasonic Waves, Fatigue Crack Monitoring, Multi-layered Structure
PACS: 43.35.Zc, 43.35.Cg, 43.35.Yb.

1. Introduction

Ensuring aircraft safety, while controlling costs, is one of the main objectives for aerospace companies. A report from Airbus [1] stresses the fact that there is limited potential for further reduction in operating costs, whereas maintenance costs can be further reduced. Maintenance costs are especially of concern for companies with ageing aircraft fleets, as the inspection manpower requirement increases over the lifespan of the aircraft [2]. The adaptation and integration of current nondestructive testing (NDT) techniques for structural health monitoring (SHM) at damage critical locations offers the potential to reduce inspection time and cost. The detection of defects in the different layers of the multi-layered aircraft structure is one of the requirements for future SHM systems [3]. Only some of the NDT techniques used today, such as ultrasonic testing (UT) and eddy currents, are considered to be mature enough for use in SHM systems [4]. Eddy current technology has been proven for the detection of fatigue cracks in aircraft structures [5, 6], but for the use in future SHM systems has the disadvantage that only local damage monitoring, e.g., at a single fastener hole, can be achieved. UT is used for the detection of fatigue cracks at fastener holes in aircraft. Under laboratory conditions and with the presence of a sealant/adhesive layer, small defects (0.025 inch EDM notch) in the 2nd layer have been detected [7]. Recently a group at QinetiQ developed a similar UT technique for 2nd layer defect detection [8]. However, 2nd layer defect detection using conventional

UT techniques can be problematic if the coupling medium (sealant) between the layers around the fastener hole is inadequate or missing [5, 8].

The possibility of fatigue crack detection at fastener holes in multi-layered structures using high frequency guided ultrasonic waves (5 MHz) has been investigated by Lindgren et al. [9]. Results from their work show that defect detection is possible, but that the detection sensitivity depends on the interface conditions between the layers. The authors also noted that high frequency guided ultrasonic waves are attenuated, if a material, such as an adhesive, is present between the metallic layers, making the monitoring of large areas difficult. According to Dalton et al. [10] potential guided wave modes used for large area inspection of double-layered metallic fuselage structures should have frequencies below 1 MHz to avoid attenuation over short distances. Therefore it is important to identify and study low frequency guided ultrasonic wave modes that are insensitive to interface conditions, but offer good sensitivity for defect detection. A flexural mode similar to the A_0 mode has energy distributed through the thickness of the multi-layered structure [11], allowing in principle defect detection in all layers. In the frequency range of interest the phase and group velocity of this A_0-like mode are insensitive to local thickness or material property variations of the sealant layer [12]. Most of the energy of the wave pulse can be propagated past areas with lack of sealant or adhesion with minimal time lag. Utilizing such a mode in the low frequency regime could overcome difficulties such as missing sealant and attenuation, offering potential benefits for the inspection and SHM of large parts consisting of multiple layers. However, low frequency guided ultrasonic waves have wavelengths significantly larger than commonly used in bulk wave UT, therefore the sensitivity for the detection of fatigue defects needs to be ascertained.

In this contribution the application of low frequency guided ultrasonic waves for the in-situ monitoring of fatigue crack growth in a multi-layered tensile specimen, consisting of two adhesively bonded aluminum alloy plate-strips, has been investigated. The guided waves in the tensile specimen were excited using piezoelectric discs and the resulting wave field was measured using a laser interferometer. Wave modes and dispersion relations for the multi-layered tensile specimen were calculated using a semi-analytical finite-element (SAFE) model. The wave propagation along the specimen was simulated from three-dimensional (3D) Finite Element (FE) calculations and compared to the experimental results. The influence of a fatigue defect in one of the metallic layers on the scattered field around the hole on the opposite layer was studied from FE simulations and experiments. Fatigue experiments were performed for 3 specimens and the amplitude of the guided ultrasonic wave field was monitored on-line at a single point using the laser interferometer.

2. Experiments

The multi-layered tensile specimens used for the fatigue testing (size: 600 mm x 40 mm x 6.2 mm) consist of two 3 mm thick aluminum alloy (2024-T6) plate-strips, adhesively bonded with a 0.2 mm thick adhesive layer (HYSOL-9394 EA) as shown in figure 1b. Three piezoelectric discs (Ferroperm Pz 27, diameter 5 mm, thickness 2 mm), permanently bonded using two-component epoxy glue 190 mm from the end of the specimen (figure 1a), were used for the excitation of the guided ultrasonic waves. Two of the discs were placed 4 mm from the respective side edges on the top layer (left in figure 1a) and one disc was placed in the centre of the opposite layer (right in figure 1a). This set up was chosen to excite a bending mode similar to the A_0 mode in a single layer plate The experimental setup shown in figure 1c was used. The excitation signal consisted of a 5 cycle, 115 kHz tone burst in a Hanning window and had narrow bandwidth. The signal was generated using an arbitrary function generator (Agilent 33220A), amplified (Krohn-Hite 7602M) and applied to the three piezoelectric discs. The excitation frequency was well below the thickness resonance of the PZT discs, which vibrate in a piston-like mode [13] and apply mostly an out-of-plane force to the specimen. For all of the experiments the velocity of the out-of-plane displacement was measured using a heterodyne laser interferometer (Polytec OFV 505/5000).

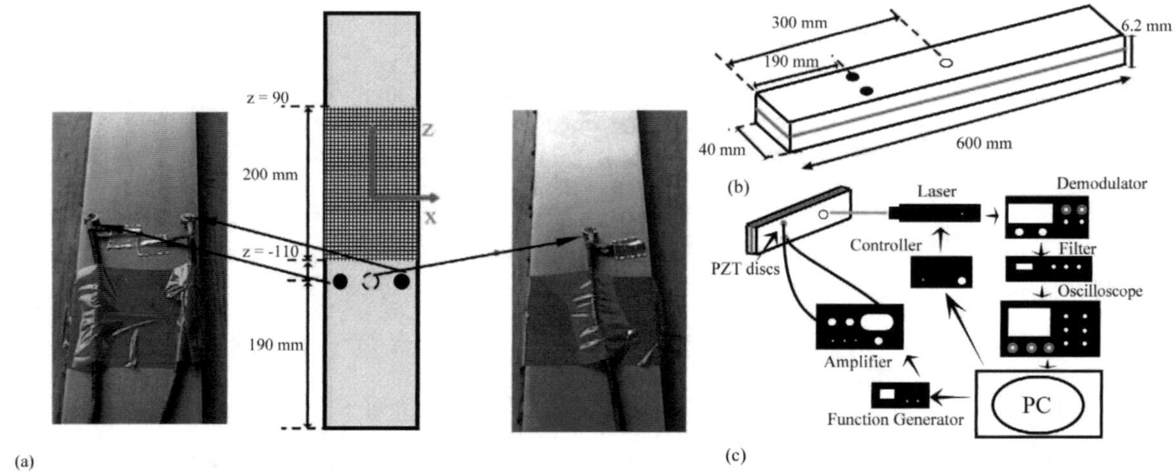

Figure 1. a) Schematic view (not to scale) of excitation setup and monitoring grid for wave propagation measurements; location of PZT discs (left: top layer; right: bottom layer); b) dimensions of multi-layered tensile specimen (600 mm x 40 mm x 6.2 mm); c) experimental setup.

The demodulated output signal obtained from the interferometer was band-pass filtered (Krohn-Hite 3988) around the centre frequency and averaged (20 times) in a digital storage oscilloscope (LeCroy 9304). Three different types of measurements were performed to study the wave propagation in the specimen, the influence of a crack on the scattered field around a fastener hole and to monitor fatigue crack growth. For the wave propagation in the multi-layered tensile specimen without a fastener hole, point wise measurements were taken on a Cartesian grid (figure 1a) every 2 mm in the z-direction and every 1 mm in the x-direction. The influence of a fatigue defect in the bottom layer on the scattered field around the fastener hole was studied. The through hole (diameter 6 mm) was drilled 300 mm from the end of the specimen (110 mm from PZT discs, figures 1b, 2a). Measurements were taken on the top, opposite layer on a radial grid around the fastener hole, every 5° in the angular direction and every 1 mm in the radial direction (red area in figure 2b, r = 4 mm to r = 12 mm). Measurements at one specimen were taken before the start of the fatigue experiments and repeated after a 4.8 mm long fatigue crack had been grown through the bottom layer. For the fatigue experiments three specimens were subjected to cyclic tensile loading in a servo-hydraulic testing machine (figure 2c). A maximum load of 30 kN was selected so that the maximum stress in the vicinity of the hole reached about 95% of the yield limit (stress concentration factor $K_t \approx 3.1$), in order to remain in the range of elastic deformation. The cyclic loading was performed with a stress ratio of $R = 0.1$ and cycling frequency of 10 Hz. An about 0.35 mm long starter notch was placed at the corner of the hole with the free surface of the bottom layer to control the crack location. After crack initiation the cyclic loading was stopped every 1000 cycles and the maximum tensile load was applied to avoid crack closure. The crack length on the bottom layer surface and depth in the fastener hole were measured optically using a microscope (specimen 1: only surface crack length). The amplitude of the guided ultrasonic wave amplitude was measured at a single point (z = -3 mm, x = 3 mm relative to the centre of the hole; shown as white dot in figure 2b).

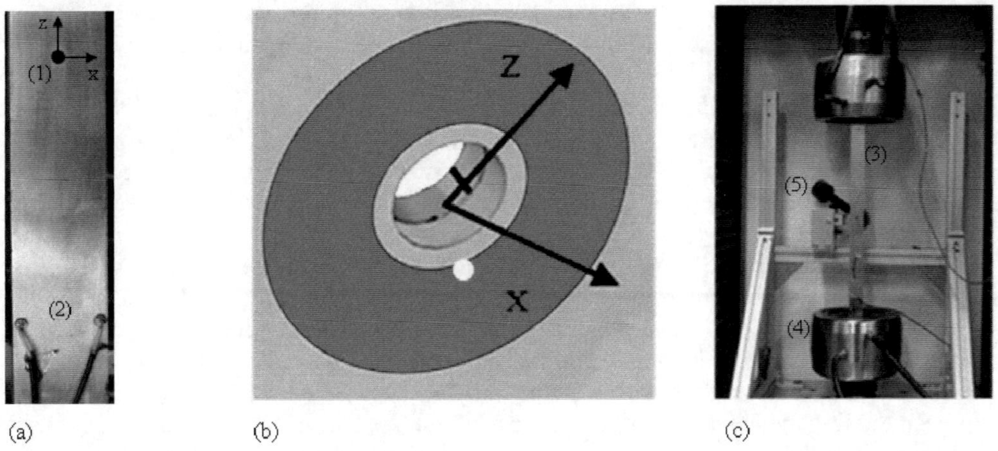

Figure 2. a) Tensile specimen with fastener hole (1) and PZT discs (2); b) schematic view of fastener hole and fatigue crack location with used coordinate system; measurement area (red area) and measurement point for single point monitoring (white dot) marked; c) fatigue setup with specimen (3) in hydraulic clamps (4) and microscope (5).

3. Guided wave propagation in tensile specimen

A semi-analytical finite-element (SAFE) code developed at the University of Bordeaux [14] was used to obtain the dispersion diagram and mode shapes for the guided ultrasonic wave propagation in the multi-layered tensile specimen. Using SAFE the cross-section of the structure is discretized using 2D elements by assuming wave propagation along the length of the specimen, reducing the 3D problem to an equivalent 2D problem. The eigenvalue problem for each frequency is solved numerically and the corresponding mode shapes and wave numbers are calculated. Results obtained from the SAFE calculations show that in the frequency range of interest, which was well below the cut-off frequencies of higher order thickness modes for an equivalent infinite multi-layered plate, the tensile specimen shows multiple modes across the width (figure 3a). The four types of modes can be classified (similar to [15]) as flexural in the thickness direction (Fx), flexural in the width direction (Fy), torsional (T) and longitudinal (L).

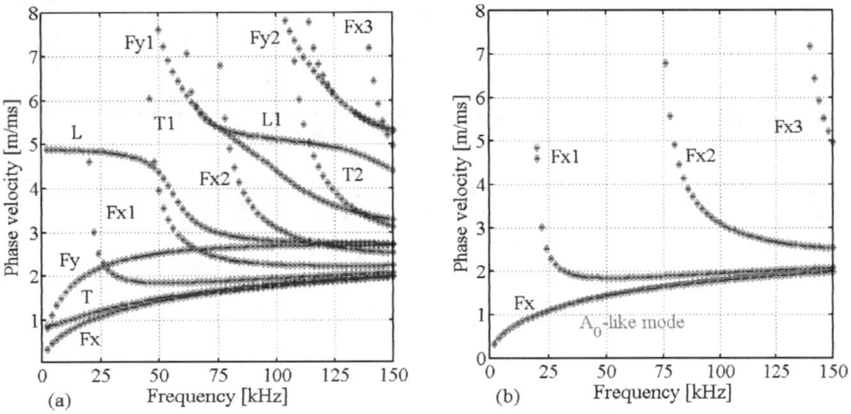

Figure 3. a) Dispersion diagram for multi-layered tensile specimen: torsional modes (T), longitudinal modes (L), flexural modes in thickness direction (Fx), and flexural modes in width direction (Fy); b) dispersion diagram showing only flexural modes in thickness direction (Fx, blue, diamonds) and A_0-like flexural mode (red, solid) for an infinite multi-layered plate (same thickness).

The excitation setup described above will mainly excite flexural modes with the largest displacement in the thickness direction (figure 3b). Higher order modes are denoted with a number, e.g., Fx1 is the first higher order flexural mode in the thickness direction with a cut-off frequency of 20 kHz. At the chosen excitation frequency of 115 kHz there are three modes propagating, namely, Fx, Fx1, and Fx2. From figure 3b it can be seen that the dispersion curve of the lowest flexural mode Fx agrees very well with the A_0-like mode in an infinite multi-layered plate [15]. For the frequency range of interest the dominant out-of-plane displacement of these bending modes does not vary significantly through the thickness (figure 4d-f). Due to the lower stiffness of the adhesive layer, the in-plane displacement in each of the metallic layers is similar to the in-plane displacement of an A_0 mode in a single layered plate with the same thickness as one of the adherents [16]. In figure 4 the mode shapes through the thickness for the three flexural modes are compared to the A_0-like bending mode. The mode shapes for the Fx mode (figure 4d) and the Fx1 mode (figure 4e) agree well with the mode shape of the A_0-like bending mode in an infinite plate, while a slight difference, mostly for the in-plane displacements, can be seen between the Fx2 mode and the A_0-like mode (figure 4f). This suggests that the flexural modes have similar sensitivity for defect detection as the A_0-like mode, which could be employed for the inspection of large multi-layered structures. From the SAFE calculations it was found that the mode shape across the width of the lowest flexural mode (Fx) is almost constant (figure 4c), while the higher order modes, Fx1 (figure 4b) and Fx2 (figure 4a) have a standing wave amplitude pattern across the width. In order to try to excite the Fx1 mode above the cut-off frequency of the Fx2 mode (75 kHz), 3 PZT discs were placed at the nodes of the Fx2 mode (x = -16 mm and x = 16 mm, see figure 4a) and at the centre of the plate strip. FE simulations were carried out using a 3D model in ABAQUS/Explicit to investigate the wave propagation in the tensile specimen. The geometry was approximated with a Cartesian mesh (plates: 0.5 mm x 0.5 mm x 0.75 mm, sealant: 0.5 mm x 0.5 mm x 0.1 mm).

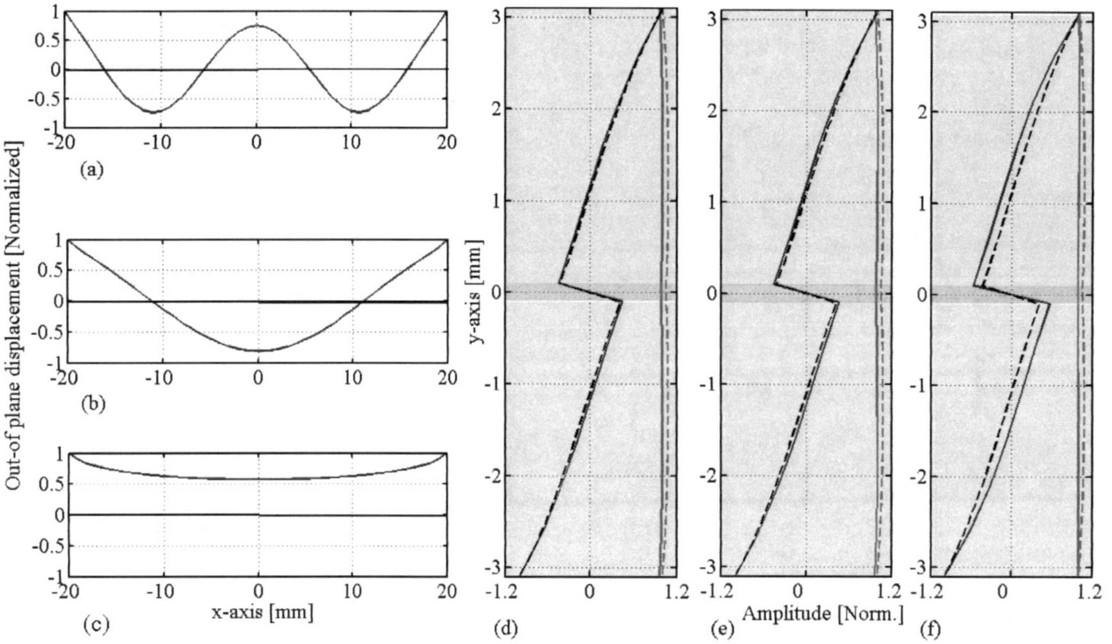

Figure 4. Mode shapes (out-of-plane amplitude) across the width: a) Fx2; b) Fx1, c) Fx; mode shapes through the thickness, out-of-plane displacement (red, solid) and in-plane displacement (blue, solid) for Fx, Fx1, and Fx2, out-of-plane displacement (green, dashed) and in-plane displacement (black, dashed) for the A_0-like bending mode: d) Fx; e) Fx1; f) Fx2; grey area: aluminum adherent; red area: adhesive layer.

The piezoelectric discs used in the experiments were modelled as out-of-plane point forces at the excitation locations and the wave field was monitored on a Cartesian grid. The excitation setup used for the experiments did not excite a single mode (Fx1) as intended, but excited multiple flexural modes. This lead to the amplitude variations seen along the width and length of the specimen in the FE simulations (figure 5a). These amplitude variations are caused by the interference between the flexural modes that can propagate at this frequency. The distance between two amplitude peaks at the centre of the specimen seen in figure 5a is about 60 mm. The measurement results in figure 5b show a similar amplitude variation to the FE simulations, but due to experimental variations the amplitude map is not symmetrical. For the study of the influence of a defect on the scattered field, the fastener hole was placed at a location of high amplitude 110 mm away from the excitation (z = 0 mm in figure 5b). The amplitude at this location is relatively constant, therefore locally around the fastener hole scattering similar to that of an A_0-like bending mode in a large plate can be expected, as the mode shapes through the thickness are similar.

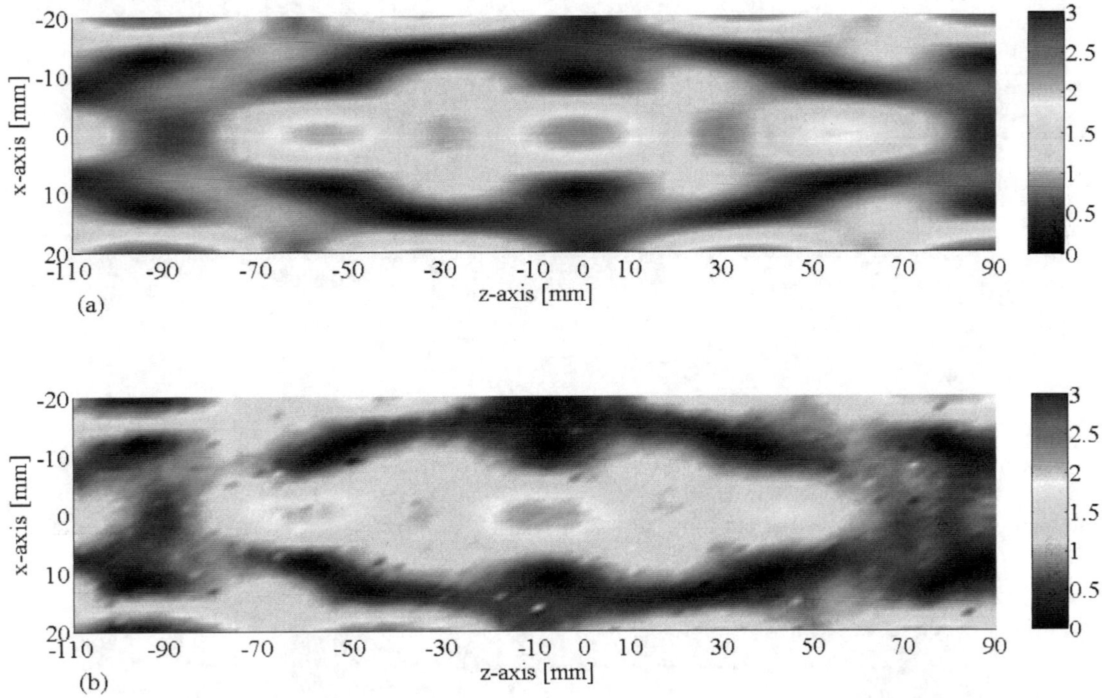

Figure 5. Amplitude variation along multi-layered tensile specimen at excitation frequency (115 kHz), monitoring grid size: 2 mm in z-direction, 1 mm in x-direction; a) FE simulations, excitation: 3 point forces at z = -110 mm; b) measurements, excitation: 3 PZT discs at z = -110 mm.

4. Influence of a fatigue crack on the scattered field

The influence of a fatigue crack in one of the layers on the scattered field around the fastener hole was investigated. The scattered field around the hole was measured before the start of the fatigue experiments and after a 4.8 mm long fatigue crack had been grown through the bottom layer at the right side of the hole. For the 3D FE simulations the defect was modeled as a zero width crack (5mm length) through the bottom layer. The fastener hole was approximated by removing elements from the Cartesian mesh. The measurement of the scattering around the hole without a defect (figure 6c) shows a similar amplitude pattern as the FE simulations (figure 6a), but is not perfectly symmetrical as the incident wave field is not symmetrical.

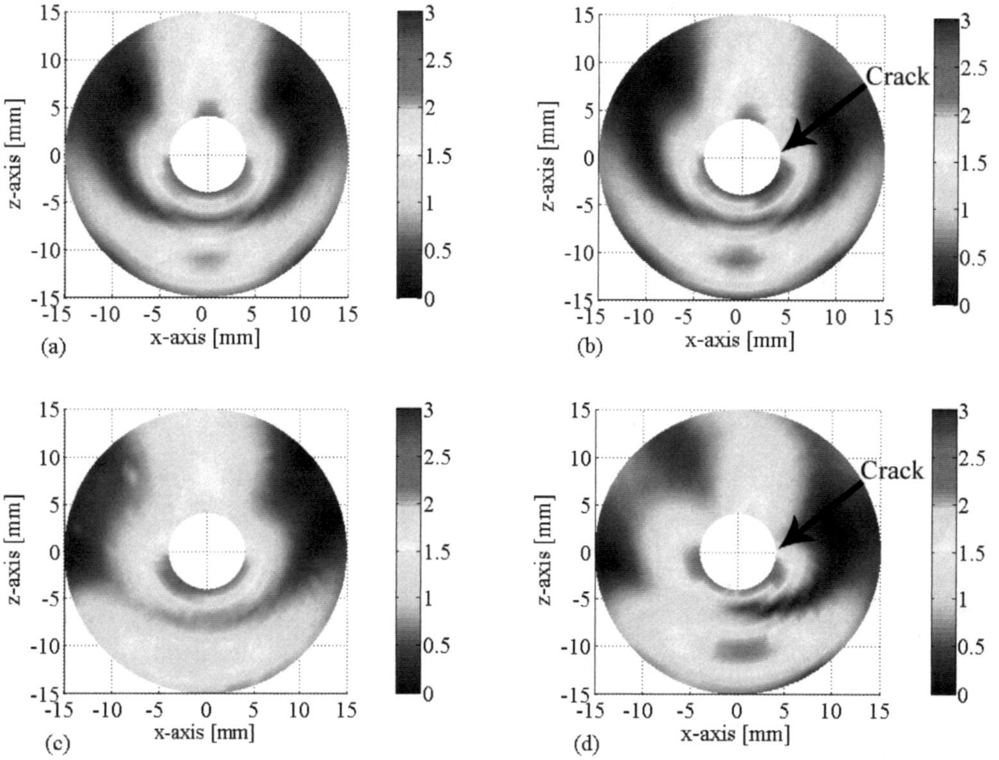

Figure 6. Amplitude of scattered field around hole (r = 3 mm) at centre frequency (115 kHz), incident wave propagating from bottom to top; a) FE simulation, no defect; b) FE simulation, 5 mm crack in bottom layer at right side of hole (z = 0 mm, x = 3 mm); c) experiment, no defect; d) experiment, 4.8 mm long fatigue crack in bottom layer at right side of hole (z = 0 mm, x = 3 mm).

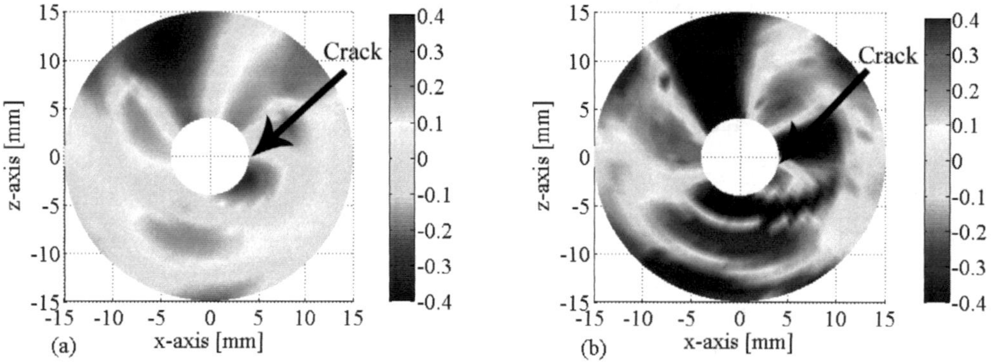

Figure 7. Difference in amplitude at centre frequency (115 kHz) between the scattered field around fastener hole (r = 3 mm) with and without fatigue crack; incident wave propagating from bottom to top; a) FE simulation, 5 mm crack in bottom layer at right side of hole (z = 0 mm, x = 3 mm); b) experiment, 4.8 mm long fatigue crack in bottom layer at right side of hole (z = 0 mm, x = 3 mm).

From the FE simulations an additional reflection of the ultrasonic wave at the fatigue crack in the bottom layer is predicted, increasing the amplitude before the defect on the right side of the fastener hole and shifting the amplitude pattern (figure 6b). This change of the scattered field can be better observed from the difference of the amplitude patterns with and without a defect (subtraction of amplitudes shown in figures 6a and 6b), shown in figure 7a. An increase in amplitude of about 30% in front of the defect due to the interference with the additional scattered wave at the defect can be seen. In the area located behind the defect a decrease of about 15% is predicted. The experimental results (figure 6d) show a similar influence of the fatigue crack on the scattered field. The scattered field is shifted due to the presence of the defect. Behind the crack an amplitude decrease of about 30% was observed (figure 7b), while the amplitude increases by about 25% in front of the crack. The measurement point for the in-situ monitoring of the fatigue crack growth was chosen in the area in front of the crack location (figure 7b at z = -3 mm, x = 3 mm), where an amplitude increase is predicted for fatigue crack growth. Another possible location to monitor changes in the amplitude would be behind the crack, where an amplitude decrease was observed previously for a similar multi-layered specimen [17].

5. Fatigue crack growth monitoring

Cyclic loading of three tensile specimens was carried out as described in section 2. During the fatigue experiments, on-line monitoring of the guided ultrasonic wave amplitude at a single point on the top surface in front of the crack (see figure 2b) was performed using the laser interferometer. From the starter notch the crack grew with quarter elliptical shape in the bottom layer. Reaching the thickness of the bottom layer (3 mm), the crack continued to grow in length as a through crack in this layer. No cracking of the top layer was observed. Figure 8a shows the optically measured crack lengths (bottom surface length and depth) against the number of cycles for the three specimens (specimen 1 only surface length). For all three specimens crack initiation occurred between 20 000 and 30 000 cycles. After 45 000 cycles the fatigue crack had grown to a length of about 5 mm on the bottom surface. Figure 8b compares the amplitude of the guided ultrasonic wave at the single measurement point against FE simulations for the variation in crack length.

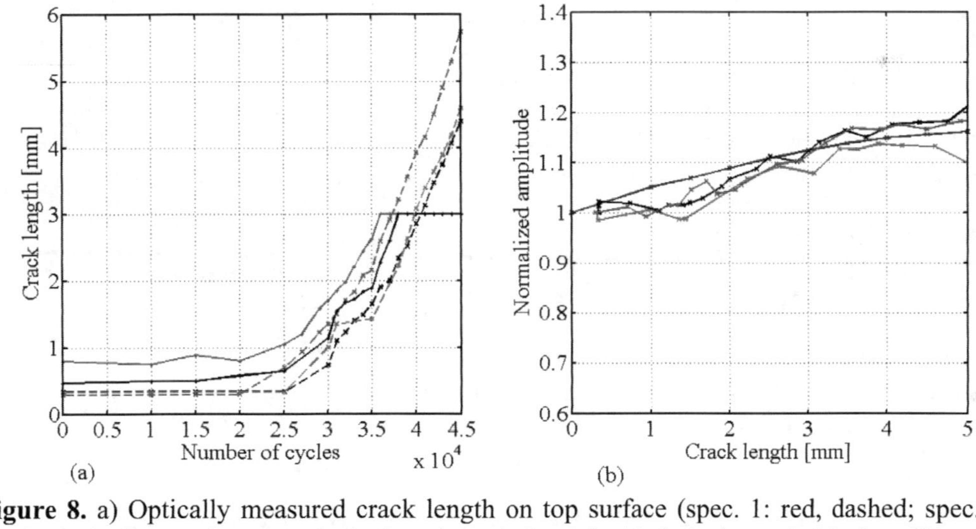

Figure 8. a) Optically measured crack length on top surface (spec. 1: red, dashed; spec. 2: black, dashed; spec. 3: green, dashed) and crack depth inside hole (spec. 2: black, solid; spec. 3: green, solid) against number of cycles; b) measured ultrasonic amplitude at single monitoring point against optically measured surface crack length, normalized with initial amplitude (FE: blue; spec. 1: red; spec. 2: black; spec. 3: green),

The monitored amplitude was normalized for each specimen with respect to the amplitude measured at the start of the fatigue experiment with no crack present. Good general agreement was found between the FE simulations and the experimental results for the 3 specimens (figure 8b). The amplitude of the guided wave at the monitoring location in front of the crack increases with increasing crack length. For a 5 mm long crack through the bottom layer the average measured amplitude increase for the three specimens is about 18% compared to a predicted 16% change for the FE simulations. This allows in principle for the detection of bottom layer fatigue cracks from guided ultrasonic wave measurements on the top layer. However, further investigations will be necessary to verify and improve the sensitivity for the detection of small cracks in real structures and to allow for remote detection without the necessity of local access to the fastener hole for the laser measurement.

6. Conclusions

The possibility of using low frequency guided ultrasonic waves for the monitoring of fatigue crack growth at fastener holes in multi-layered tensile specimens was investigated. The wave propagation in the multi-layered tensile specimen was predicted from SAFE and 3D FE calculations. Amplitude variations along and across the specimen were observed, indicating multiple flexural modes (across the width) propagating at the frequency of interest. The thickness mode shapes of the flexural modes show good agreement with an A_0-like bending mode in an infinite multi-layered plate, predicting similar sensitivity for defect detection. The amplitude of the scattered field around the hole with and without a defect was recorded on the top layer experimentally and from 3D FE simulations, showing a similar amplitude pattern for the undamaged case. The influence of a fatigue crack in the bottom layer on the scattered field was investigated. The scattered field obtained from the FE calculations for a defect showed good qualitative agreement with measurements, predicting a shift of the scattered field and an increase in amplitude in front of the defect. The amplitude at one point close to the defect was monitored during cyclic loading of three tensile specimens. Results from the on-line monitoring of fatigue crack growth and FE simulations showed a change in amplitude of the guided ultrasonic wave field with increasing crack length. The measured ultrasonic wave amplitude increase for the three specimens shows some variation but agrees well with the predicted increase from FE simulations. The measurement of the guided wave amplitude change allows in principle for the detection of fatigue cracks from the laser measurements close to the fastener hole. The sensitivity of the technique employing low frequency guided ultrasonic waves for the detection of fatigue cracks in tensile specimens has been investigated in this contribution. In order to employ this technique for future aircraft SHM applications, a number of steps would have to be implemented to ascertain the realistic application, reliability, and robustness of this technique for large multi-layered structures.

Acknowledgements
This work was supported by grant EP/D065011/1 from the UK Engineering and Physical Sciences Research Council (EPSRC).

References

[1] Speckmann H and Henrich R 2004 Structural health monitoring - overview on technologies under development *Proceedings of the 16th World Conference on NDT, Montreal, Canada*
[2] Sampath S G 1996 Aging combat aircraft fleets - long term applications *NATO AGARD LS-206, Neuilly-sur-Seine, France*
[3] Smith C D 1996 Federal Aviation Administration Aircraft Inspection Research and Development Programs *Nondestructive Evaluation of Aging Aircraft, Airports and Aerospace Hardware ed. by Rempt R D and Broz A L SPIE Conference Proceedings* **2945** 200-09
[4] Boller C 2000 Next generation structural health monitoring and its integration into aircraft design *Int. J. Syst. Sci.* **31** 1333-49

[5] Lindgren E, Knopp J S, Aldrin J C, Steffes G J and Buynak C F 2007 Aging aircraft NDE: capabilities, challenges and opportunities *in Review of Progress in Quantitative NDE* **26** *ed. by Thompson D O and Chimenti D E AIP Conference Proceedings* **894** 1731-38

[6] LaRiviere S G and Thompson J 1998 Development of reliable NDI procedures for airframe inspection *NATO RT0 MP-10- Paper 7, Brussels, Belgium*

[7] Lindgren E, Judd D, Concordia M, Mandeville J, Aldrin J C, Spencer F, Fritz D, Pratt E, Waldbusser R and Mullis R T 2004 Validation and deployment of automated ultrasonic inspections for the C-130 center wing *ASIP Conference Savanna, Georgia*

[8] Smith R A, Edgar D, Jones L D and Percivall D M 2005 An ultrasonic solution for second-layer crack detection *Insight* **47** 81-84

[9] Lindgren E, Aldrin J C, Jata K, Scholes B and Knopp J 2007 Ultrasonic plate waves for fatigue crack detection in multi-layered metallic structures *Health Monitoring of Structural and Biological Systems ed. by Kundu T SPIE Conference Proceedings* **6532** 53207

[10] Dalton R P, Cawley P and Lowe M J S 2001 The potential of guided waves for monitoring large areas of metallic aircraft fuselage structure *J. Nondestr. Eval.* **20** 29-46

[11] Rokhlin S I 1991 Lamb wave interaction with lap-shear adhesive joints: theory and experiment *J. Acoust. Soc. Am.* **89** 2758-65

[12] Lowe M J S and Cawley P 1994 The applicability of plate wave techniques for the inspection of adhesive and diffusion bonded joints *J. Nondestr. Eval.* **13** 185-200

[13] Guo N and Cawley P 1991 Transient response of piezoelecric discs to applied voltage pulses *Ultrasonics* **29** 208-17

[14] Predoi M V, Castaings M, Hosten B and Bacon C 2007 Wave propagation along transversely periodic structures *J. Acoust. Soc. Am.* **121** 1935-44

[15] Mukdadi O M, Dessai Y M, Datta S K, Shah A H and Niklasson A J 2002 Elastic guided waves in a layered plate with rectangular cross section *J. Acoust. Soc. Am.* **112** 1766-79

[16] Seifried R, Jacobs L J and Qu J M 2002 Propagation of guided waves in adhesive bonded components *NDT & E Int.* **35** 317-28

[17] Kostson E and Fromme P 2009 Defect detection in multi-layered structures using guided ultrasonic waves *in Review of Progress in Quantitative NDE* **28** *ed. by Thompson D O and Chimenti D E AIP Conference Proceedings* **1096** 209-16.

Quasi-static elastography comparison of hyaline cartilage structures

A J McCredie, E Stride and N Saffari
Mechanical Engineering, University College London, Torrington Place, London, WC1E 7JE

Abstract. Joint cartilage, a load bearing structure in mammals, has only limited ability for regeneration after damage. For tissue engineers to design functional constructs, better understanding of the properties of healthy tissue is required. Joint cartilage is a specialised structure of hyaline cartilage; a poroviscoelastic solid containing fibril matrix reinforcements. Healthy joint cartilage is layered, which is thought to be important for correct tissue function. However, the behaviour of each layer during loading is poorly understood. Ultrasound elastography provides access to depth-dependent information in real-time for a sample during loading. A 15 MHz focussed transducer provided details from scatterers within a small fixed region in each sample. Quasi-static loading was applied to cartilage samples while ultrasonic signals before and during compressions were recorded. Ultrasonic signals were processed to provide time-shift profiles using a sum-squared difference method and cross-correlation. Two structures of hyaline cartilage have been tested ultrasonically and mechanically to determine method suitability for monitoring internal deformation differences under load and the effect of the layers on the global mechanical material behaviour. Results show differences in both the global mechanical properties and the ultrasonically tested strain distributions between the two structures tested. It was concluded that these differences are caused primarily by the fibril orientations.

1. Introduction

Cartilage is a structural load bearing tissue which is found throughout the bodies of mammals. Articular joint cartilage is subject to rigorous cyclic loading patterns in everyday use, but is prone to damage under sudden shock loading conditions (Verteramo and Seedhom 2007). Once damaged, articular cartilage has limited ability for self-repair (Klein et al. 2003; Yu et al. 1997) and consequently is a target for regenerative medicine (Yu et al. 1997; Klein et al. 2003; Ma et al. 1995; Grellmann et al. 2006). Articular cartilage has a fibril-reinforced layered poroviscoelastic structure (Fung 1993; Maroudas 1979; Mow et al. 1991) and as such has poorly characterised mechanical behaviour. In order to provide a prescriptive specification of the tissue properties required for engineered tissue, the contributions of the various structural features need to be determined.

Articular cartilage is a specialised structure of hyaline cartilage (Maroudas 1979). Hyaline cartilage consists of a porous, collagen fibril reinforced proteoglycan matrix containing small clusters of cells and an interstitial fluid (Fung 1993). The layered structure of cartilage tissue is shown in figure 1 (Mow et al. 1991). The direction of the collagen fibrils within the layers is a primary determinant of their mechanical behaviour. In the superficial zone the fibrils are in parallel alignment with the tissue surface, in the middle zone the fibrils are randomly aligned and in the deep zone the fibrils are perpendicular to the subchondral bone. In the superficial and deep zones the collagen fibrils exist in a higher density than in the middles zone Mow and Guo 2002. The global (depth-independent) mechanical behaviour of the tissue is determined by an interaction of the matrix porosity, the intrinsic matrix viscoelasticity and the fibril reinforcement (Mak 1986; DiSilvestro and Suh 2002; Haider and Schugart 2006). Under loading conditions the fluid in the pores is pressurised and subsequently flows down the pressure gradient. This pressurisation and flow coupled with a dependency of permeability on strain (Mansour and Mow 1976) causes the properties to be strain and strain rate dependent (Park et al. 2008). The fibrils also exhibit tension-compression non-linearity (Soulhat et al. 1999) and strain dependence (Fortin et al. 2000). The layered structure and properties of articular cartilage are essential for correct tissue function (Chen et al. 2001; Klein et al. 2003).

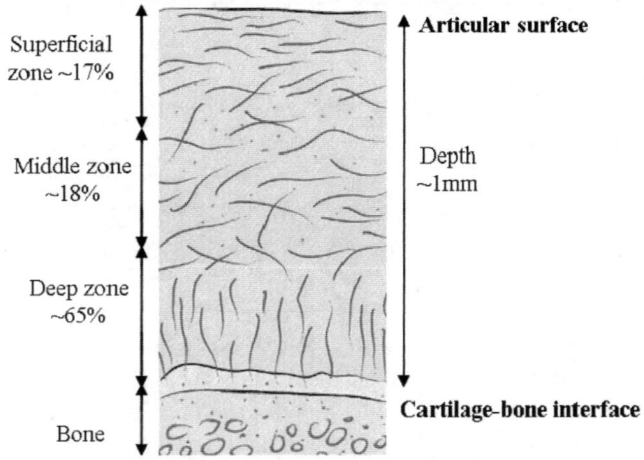

Figure 1: Illustration of the cross-section of articular cartilage. The fibril alignment is highlighted. Taken from Mow et al. (1991).

Previous investigations into cartilage behaviour have included both ultrasonic (Berry et al. 2006; DiSilvestro et al. 2001; Toyras et al. 1999; Zheng et al. 2002) and mechanical (Mak 1986; Fortin et al. 2000; Buschmann et al. 1998; Park et al. 2008) testing methods. The collagen fibrils have been seen to affect the backscattering characteristics of the ultrasonic signal (Laasanen et al. 2005). Measurements of attenuation and reflection characteristics have both been demonstrated to show potential for diagnosis of cartilage degeneration (Nieminen et al. 2004; Toyras et al. 1999), but neither of these parameters allows direct mechanical property measurements. Characterisation of the global mechanical behaviour of cartilage has been extensive (DiSilvestro et al. 2001; Toyras et al. 1999; Zheng et al. 2002; Mak 1986; Fortin et al. 2000; Buschmann et al. 1998; Park et al. 2008), but although the depth-dependent mechanical properties contribute to the global properties, the importance of the separate layers also needs to be considered for tissue engineering purposes.

Elastography is an ultrasonic technique initially developed for the detection of cancerous tumours (Ophir et al. 1991). The elastography procedure involves acquiring ultrasonic backscattered signals from a material before and after the application of a strain. The shifts observed in the arrival times of signals from internal scatterers are used to determine localised tissue strains under the assumption of a constant sound speed in the material. The strains in the tissue are inversely proportional to the elastic moduli and hence this provides a method for the characterisation of depth-dependent moduli, as long as sufficient information regarding the material stresses is known. The feasibility of material characterisation of poroviscoelastic materials (such as cartilage) via elastography has been investigated (Cohn et al. 1997a; Cohn et al. 1997b; Righetti et al. 2005; Zheng et al. 2002; Zheng et al. 2005). The main limitations of these previous cartilage investigations were the boundary conditions from the method of strain application and the sample orientations. In the studies by Zheng et al. the ultrasonic data were obtained using a 50 MHz transducer, the beam from which was significantly attenuated before reaching the furthest layers of the material (Zheng et al.). The work reported here is an improvement of the methodology first reported by Zheng et al. The nasal cartilage samples used in the present study were approximately double (3.93±0.55mm) the thickness of the articular samples (1.57±0.30mm), which would cause an unacceptable degree of attenuation, so a lower frequency transducer was selected. The other key difference is that in this present study the deflection of the load cell was quantified using a linear variable differential transformer (LVDT); the level of load cell deformation (approximately 0.0016mm/N) is only one order of magnitude smaller than the displacement applied to the cartilage, so are therefore relevant to the results. The experiments performed by Zheng et al. included only a linear approximation for the load cell deflection.

This current work is a preliminary study for future work involving cyclic loading protocols to link the layered structure of cartilage to the behaviour in more physiologically relevant conditions.

2. Materials and Methods

In this study a quasi-static elastography method was used to compare the depth-dependent mechanical behaviour of two types of hyaline cartilage. These were bovine articular cartilage and porcine nasal cartilage. Nasal cartilage is a hyaline cartilage with a non-layered structure (Fung 1993) whereas articular cartilage has three layers. The global mechanical properties were measured simultaneously with the acquisition of the ultrasound A-scans for elastography. Under constant strain the tissues undergo stress-relaxation, causing a reduction in the stress required to maintain the global strain (Mak 1986). The localised shift behaviour within the two types of cartilage provides information regarding the contribution of the various zones of articular cartilage to the overall tissue deformation.

Samples: Cartilage samples were obtained from a local abattoir (Blixies Farm, Ranks Green, Essex, UK) within 4 hours of slaughter, excised from surrounding tissue and prepared into 8mm diameter discs using biopsy punches. Once prepared the samples were stored in phosphate buffered saline solution (PBS) prior to testing. Ten samples for each test of each cartilage type were prepared. As shown in figure 1, the axial orientation of the articular cartilage samples was important in the tests. Nasal cartilage sample discs were axial orientation independent due to the homogeneity of the structure.

Mechanical testing: A schematic of the apparatus is shown in figure 2. The mechanical test equipment consisted of an actuator (Physik Instrumente, M-227) in series with a load cell (Interface, SM-50N) within a rigid frame. To quantify the load cell deformations a LVDT (Omega, GP911-0.5-S) was placed in parallel with the load cell. The data from the LVDT was used to determine the actual strain applied to the samples at any given point.

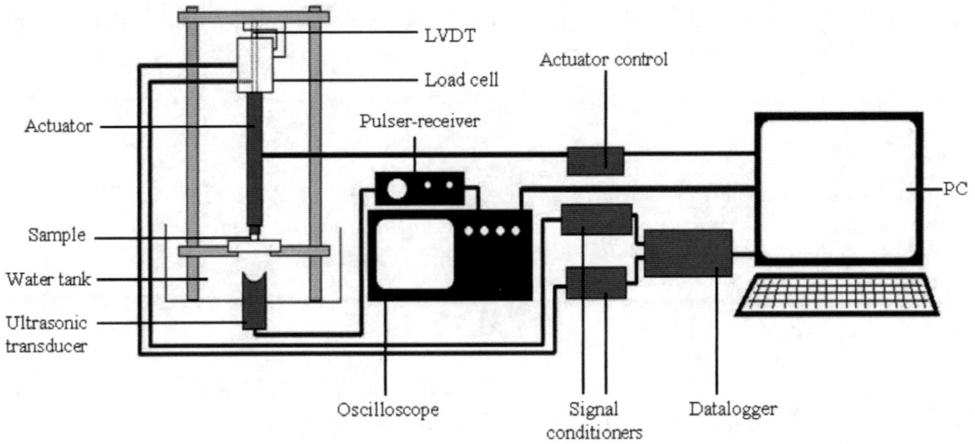

Figure 2: Schematic of the elastography equipment layout

Strains of 2, 4, 6 and 8% were applied to the top surface of the samples via the actuator and held constant for 20 minutes. The bottom surface of the samples (cartilage-bone interface of articular samples) rested on the rigid fixed surface of the Perspex delay-line. The load cell continuously recorded data of the force required to maintain the strain applied to the sample. The LVDT continuously recorded the small deformations across the load cell.

A pre-load of 3N was applied to each of the samples prior to the strain application to ensure effective contact of the samples and the compression surfaces. The application of pre-loads is standard practice in the mechanical testing of cartilage (Fortin et al. 2003; Julkunen et al. 2008). The pre-loads were maintained for approximately 5 minutes until the load cell reading reduced to zero (i.e. when no additional force was required to deform the samples). All tests were conducted with the samples fully submerged in a tank containing PBS at room temperature (22.7 ± 1.4°C).

Ultrasonic testing: The 15 MHz focused ultrasound transducer (Panametrics, V319) was inserted through the base of the tank. The lower plate of the rigid frame had a hole cut in the centre as an acoustic window and a Perspex delay-line was used for the sample to rest on. The delay-line thickness was chosen so that the ultrasound focussed into a column in the samples. Perspex was selected due to the absence of grain boundaries that would be found in a metal. The signal was transmitted and received by the same transducer using a pulser-receiver (Panametrics 5055PR). A-scans were automatically acquired at 3 second intervals throughout the test using the oscilloscope (LeCroy 9310M) controlled by an in house LabVIEW (National Instruments Corporation 2008) program.

Speed of sound values were obtained using the time of flight method demonstrated in equation 1, where a is the sample thickness, v is the sound speed and t_1 and t_2 are the arrival times of echoes from the two sample surfaces.

$$v = \frac{2a}{t_2 - t_1} \tag{1}$$

The pre-strain A-scans were compared to post-strain A-scans at three different time intervals: immediately post-strain application and 10 and 20 minutes after the strain was applied.

Analysis: Calculations of global elastic properties were made by defining the stress (σ) from initial sample dimensions and the load cell data. Global applied strain values were corrected with the LVDT data to obtain the actual strain (ε). Young's modulus, *E*, was calculated from these values as shown in equation (2):

$$E = \frac{\sigma}{\varepsilon} \quad (2)$$

Ultrasonic A-scans pre- and post-straining were processed in Matlab (The MathWorks 2008) by splitting the signals into windows and applying the sum squared difference (SSD) algorithm. The equation describing the SSD algorithm is as follows:

$$R_{SSD}(\tau) = \int_{-T/2}^{T/2} \left(s_1(t) - s_2(t+\tau)\right)^2 dt \quad (3)$$

Over a window of $\pm T/2$, signal s_2 is moved across signal s_1 in increments of τ. When the value of R_{SSD} is found to be a minimum, the corresponding value of τ is the shift between the two signals.

The windows which were one tenth of the signal length (corresponding to approximately one tenth of the sample depth) were applied to both the pre- and post-compression signals. The post-strain signal is moved step-wise through the pre-strain signal applying the SSD algorithm at each step. The SSD integral values for each step are calculated. The minimum value of the integral indicates the best match between pre- and post-strain signal shape and the corresponding step value is recorded. This step value is then converted into a time shift using the known sampling interval. The time shifts are then plotted in terms of their representative depth within the tissue. With accurate measurements of tissue thickness before and after the compression, these time shift values could be converted into displacement measurements in terms of actual tissue depth. Each window overlapped the preceding window by one third.

To ensure no erroneous data was plotted on the shift traces from portions of the signal with very little correlation, the correlation coefficient of the two signals at the step-position corresponding to the minimum SSD was also calculated. The correlation coefficient (*C*) is a measure of the degree of similarity between two signals. Whereas the SSD detects the shift at which the minimum difference occurs, the correlation coefficient provides a normalised value of the sum of the window contents for comparison. The individual points in the pre- or post-compressional signal, for example s_{1_i} where i = 1,2…(number of points in the window, (n)) is compared to the mean of the window ($\bar{s}_1$), combined with the same values in the comparison window and divided by the corresponding standard deviations (σ_{s_1}) and normalised to n:

$$C_{s_1 s_2} = \frac{\sum\left((s_{1_i} - \bar{s}_1)(s_{2_i} - \bar{s}_2)\right)}{(n-1)\sigma_{s_1}\sigma_{s_2}} \quad (4)$$

The value returned is between 0 and 1, where 0 indicates no correlation between the signals and 1 indicates identical signals. A threshold value is set for the minimum acceptable correlation coefficient. Shifts corresponding with correlation values below the threshold are not included in the graphical output of the shift profile.

3. Results

Mechanical results: Figures 3 (a) and (b) show the mechanical results from the study. The results were averaged over the 10 samples for each strain level in each tissue type; the error bars show the standard deviation at each point. Values are plotted at 15 second intervals. Figures 3 (a) and (b) compare the apparent elastic moduli throughout the full 20 minute test periods of the nasal (a) and the articular (b) samples at each of the strain levels. The apparent elastic moduli for articular cartilage appeared to be dependent on the strain applied. This was not the case for nasal cartilage samples.

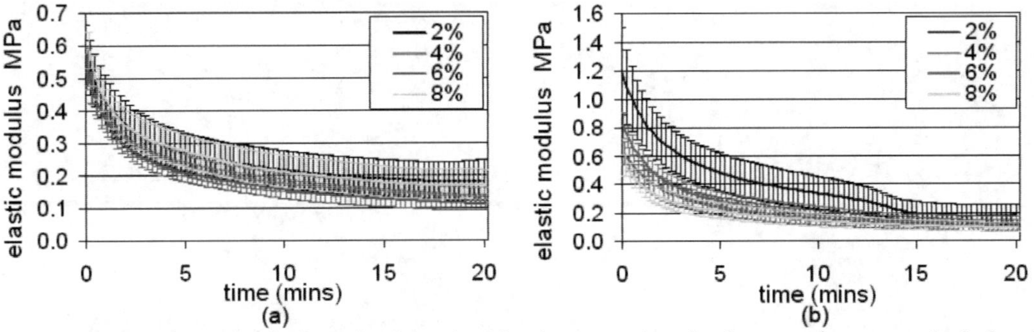

Figure 3: Mechanical results from the 20 minute quasi-static elastography tests at 2, 4, 6 and 8% applied strain. (a) Apparent elastic moduli for nasal cartilage samples; (b) Apparent elastic moduli for articular cartilage samples

Ultrasonic results: The sound speed of the cartilage was calculated using the time-of-flight method (equation 1) and the strains known to be applied to the material. The sound speeds were found to be 1920 ± 95 m/s and 1690 ± 109 m/s prior to strain application for the articular and nasal samples respectively. The sound speeds were also calculated immediately post-strain application and at 10 and 20 minutes post-strain. It was found that for both the nasal and the articular cartilage the speed of sound propagation increased under strain, however not as a function of the strain directly but rather as a function of the time the strain had been applied. Nasal cartilage sound speed increased by 0.6% ± 0.05%, 1.3% ± 0.05% and 1.9% ± 0.05% compared to unstrained cartilage at the instant of strain application and at 10 and 20 minutes respectively. Articular cartilage sound speed increased by 1.2% ± 0.05%, 3.0% ± 0.05% and 4.0% ± 0.05% compared to unstrained cartilage at the instant of strain application and at 10 and 20 minutes respectively.

Figures 4 (a)-(d) show the nasal strain maps at (a) 2%, (b) 4%, (c) 6% and (d) 8% global strains. Internal strains at the instant of load application and 10 minutes and 20 minutes later are shown with lines of best fit plotted through the data. A reference line is plotted in each figure for the strains expected in a homogeneous linear elastic material. 'Initial tissue depth', as displayed on the x-axis of figures 4 and 5, refers to the linear distance in the unstrained tissue from the bottom of the sample. The strain maps for the nasal cartilage samples show approximately linear trends at gradients lower than the reference lines. The coefficient of determination (R^2) values were recorded as a measure of how closely the trend lines fitted the data. All the R^2 values were above 0.8, which denotes a reasonably close fit, with the exception of the value for 6% strain at 10 minutes which was 0.72. Figures 5 (a)-(d) show the corresponding articular cartilage strain maps at each of the applied global strains. Lines of best fit have not been fitted to this data; only the reference lines are shown as above. The strain maps from the articular cartilage samples show regions of different gradients which deviate from the reference lines.

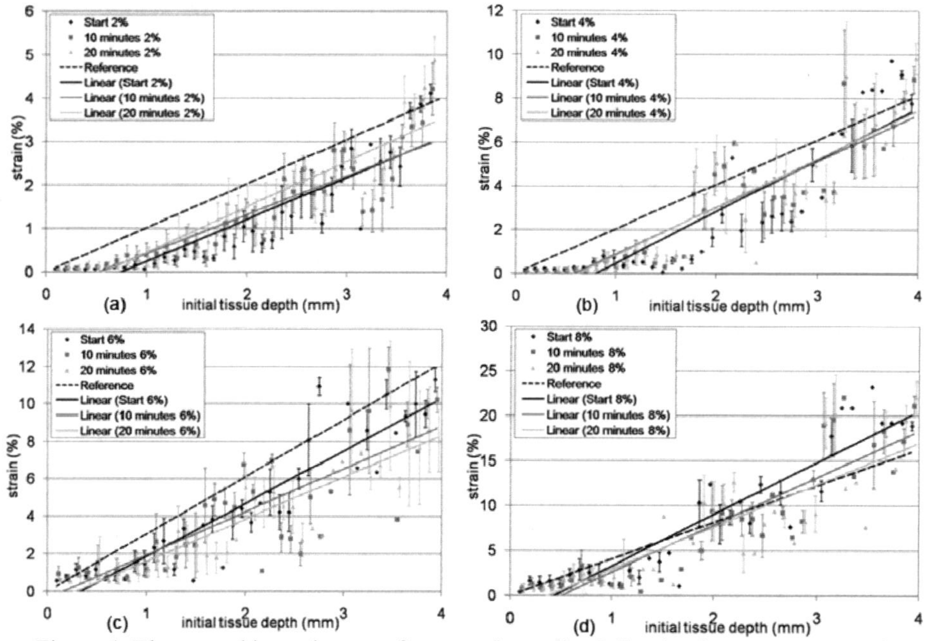

Figure 4: Elastographic strain maps from nasal samples following instantaneous strain application and after the initial strain has been held constant for 10 and 20 minutes (a) 2% applied strain; (b) 4% applied strain; (c) 6% applied strain; (d) 8% applied strain.

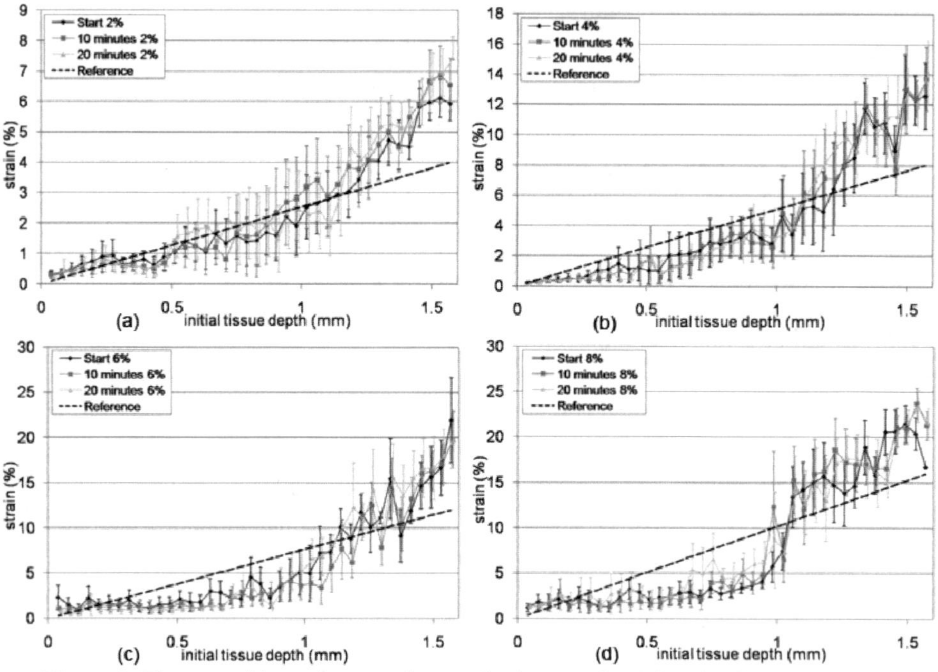

Figure 5: Elastographic strain maps from articular samples following instantaneous strain application and after the initial strain has been held constant for 10 and 20 minutes (a) 2% applied strain; (b) 4% applied strain; (c) 6% applied strain; (d) 8% applied strain.

4. Discussion

The aim of this study was to investigate the depth-dependent deformation of two different cartilage types using ultrasound elastography. Ultrasound elastography was selected as a non-destructive technique to simultaneously collect global and depth-dependent elastic properties of the tissues. The two cartilage types were compared to demonstrate which behaviours of articular cartilage are likely to be directly linked to the three-layer structure of the material.

The mechanical results demonstrated a difference between nasal and articular cartilage in terms of stress relaxation and instantaneous modulus characteristics. The articular cartilage was found to be nonlinear with the instantaneous moduli being a function of the applied strain. The instantaneous modulus was greatest at 2% compression (1.19±0.3MPa), and then decreased with applied strain (0.82±0.09MPa at 4%, 0.62±0.08MPa at 6% and 0.53±0.07MPa at 8% strain). The instantaneous moduli of the nasal cartilage samples appeared independent of the applied strain (average 0.56±0.06MPa, maximum 0.59±0.06MPa and minimum 0.52±0.05MPa). The equilibrium moduli of all the cartilage types at all applied strains were similar (approximately 0.15±0.08MPa for the nasal cartilage and 0.13±0.09MPa for the articular cartilage). Elastographically, the strain maps from the nasal cartilage samples were roughly linear and the articular cartilage strain maps showed evidence of separate behaviour in different regions. The stress-relaxation behaviour of articular cartilage was not obviously depth-dependent as small increases were seen in the strains at 10 and 20 minutes, in all measured depths. These increases were found at each level of global applied strain. The nasal cartilage data decorrelated between pre- and post-strain far more readily than the articular data. Consequently there are some places in the nasal strain maps where either points have been omitted or where strains anomalous to the trend lines have been plotted as only one or two values were available to plot from the ten averaged samples. The pre- and post-strain correlation for the articular data was better with at least 5 data points to average at any tissue depth.

The speeds of sounds measured in the two cartilage types differed. The bovine articular cartilage sound speed also varied significantly from the values found in literature for the same material (1636 ± 25m/s (Patil et al. 2004)). There was a high degree of scatter in both the mechanical and the ultrasonic data, as demonstrated by the error bars in figures 3, 4 and 5. This scatter is likely to have been caused primarily by the biological variation between individual samples, but variation in the ambient temperature would have also affected the material behaviour particularly with respect to the speed of ultrasound propagation.

The ultrasonic results from the nasal tissue showed a scattered linear strain map, with no clear individual regions. These linear strains were seen to increase with the level of applied strain. At all global strains the nasal cartilage strain maps showed a slight increase in measured strain at all tissue depths at both 15 and 20 minutes. The strain maps from the articular cartilage demonstrated regions of similar elastic moduli. The approximate thicknesses of these regions are 0.5mm for the deeper zone (31%), 0.7mm for the transitional zone (44%) and 0.4mm for the superficial zone (25%). The steeper gradient from the top regions of the articular cartilage strain maps and high peak strain values reflect the lower elastic modulus of the tissue in this region (Zheng et al. 2002). The regions corresponding to the transitional and deeper zones have less steep gradients as a result of higher elastic moduli in these regions. The region of the strain maps roughly corresponding to the transitional zone of the articular cartilage closely resembles the appearance of the whole of the nasal strain maps. Articular cartilage and nasal cartilage have similar compositions with the main difference being the collagen fibril orientation. It is likely therefore that the differences seen in the strain maps, between the two cartilage types, can mainly be attributed to the fibril arrangements.

The instantaneous mechanical behaviour of the cartilage tissues demonstrates a clear difference between the two cartilage structures investigated. The global elastic properties of articular cartilage suggest a link between the layered structure and the nonlinearity of the response to an instantaneous load. The equilibrium elastic properties of both cartilage types were approximately equal, suggesting this is not dependent on the collagen fibril orientation of the cartilage. The variation in elastic properties in the articular cartilage appears to be physiologically significant, with greatest levels of deformation permitted far away from the bone.

The diameter of the samples with respect to the measurement area covered by the beam was considered to be sufficiently large for edge-effects to be disregarded in the ultrasonic data. However, it is likely that the time taken for all the cartilage samples to reach equilibrium was reduced compared to the period *in vivo* as the fluid could move freely out of the sample edges. The speed of sound in the articular and the nasal cartilage were respectively assumed to be constant through the thickness of the tissues and independent of strain. The assumption of depth-independence for sound speed in articular cartilage is likely to be a source of error, as Agemura et al. (Agemura et al. 1990) reported a variation of approximately 5% from the superficial to the deep zones. Greater accuracy could be obtained for the ultrasonic data by adopting methods for tighter control of the ambient temperatures.

5. Conclusion

Ultrasound elastography has been used for the first time to compare the depth-dependent mechanical behaviour of two different hyaline cartilage structures. Articular cartilage demonstrated strain dependence in instantaneous elastic moduli, which was absent in the behaviour of nasal cartilage. The elastography strain maps revealed evidence of depth-dependent elastic moduli for articular cartilage while indicating a homogeneous elastic modulus for nasal. As both materials are poroviscoelastic hyaline cartilages, this result implies a direct link between the layered collagen fibril alignment and the non-linear elastic response of articular cartilage tissue. This link between the layered structure and function is not trivial and is a challenge to be met in the future for tissue engineering of articular cartilage.

Acknowledgements

Dr Vehid Salih (UCL) for assistance with the sample handling post-collection.

Reference List

Agemura, D. H., Obrien, W. D., Olerud, J. E., Chun, L. E., and Eyre, D. E. 1990. Ultrasonic Propagation Properties of Articular-Cartilage at 100 Mhz. *Journal of the Acoustical Society of America* **87** [4], 1786-1791

Berry, G. P., Bamber, J. C., Armstrong, C. G., Miller, N. R., and Barbone, P. E. 2006. Towards an acoustic model-based poroelastic imaging method: I. Theoretical foundation. *Ultrasound in Medicine and Biology* **32** [4], 547-567

Buschmann, M. D., Soulhat, J., Shirazi-Adl, A., Jurvelin, J. S., and Hunziker, E. B. 1998. Confined compression of articular cartilage: Linearity in ramp and sinusoidal tests and the importance of interdigitation and incomplete confinement. *Journal of Biomechanics* **31** [2], 171-178

Chen, A. C., Bae, W. C., Schinagl, R. M., and Sah, R. L. 2001. Depth- and strain-dependent mechanical and electromechanical properties of full-thickness bovine articular cartilage in confined compression. *Journal of Biomechanics* **34** [1], 1-12

Cohn, N. A., Emelianov, S. Y., Lubinski, M. A., and ODonnell, M. 1997a. An elasticity microscope .1. Methods. *Ieee Transactions on Ultrasonics Ferroelectrics and Frequency Control* **44** [6], 1304-1319

Cohn, N. A., Emelianov, S. Y., and ODonnell, M. 1997b. An elasticity microscope .2. Experimental results. *Ieee Transactions on Ultrasonics Ferroelectrics and Frequency Control* **44** [6], 1320-1331

DiSilvestro, M. R. and Suh, J. K. F. 2002. Biphasic poroviscoelastic characteristics of proteoglycan-depleted articular cartilage: Simulation of degeneration. *Annals of Biomedical Engineering* **30** [6], 792-800

DiSilvestro, M. R., Zhu, Q. L., Wong, M., Jurvelin, J. S., and Suh, J. K. F. 2001. Biphasic poroviscoelastic simulation of the unconfined compression of articular cartilage: I - Simultaneous prediction of reaction force and lateral displacement. *Journal of Biomechanical Engineering-Transactions of the Asme* **123** [2], 191-197

Fortin, M., Buschmann, M. D., Bertrand, M. J., Foster, F. S., and Ophir, J. 2003. Dynamic measurement of internal solid displacement in articular cartilage using ultrasound backscatter. *Journal of Biomechanics* **36** [3], 443-447

Fortin, M., Soulhat, J., Shirazi-Adl, A., Hunziker, E. B., and Buschmann, M. D. 2000. Unconfined compression of articular cartilage: Nonlinear behavior and comparison with a fibril-reinforced biphasic model. *Journal of Biomechanical Engineering-Transactions of the Asme* **122** [2], 189-195

Fung, Y. C. 1993. Biomechanics: Mechanical Properties of Living Tissue. 2nd Ed. New York, Springer.

Grellmann, W., Berghaus, A., Haberland, E.-J., Jamali, Y., Holweg, K., Reincke, K., and Bierogel, C. 2006. Determination of strength and deformation behavior of human cartilage for the definition of significant parameters. *Journal of Biomedical Materials Research Part A* **78A** [1], 168-174

Haider, M. A. and Schugart, R. C. 2006. A numerical method for the continuous spectrum biphasic poroviscoelastic model of articular cartilage. *Journal of Biomechanics* **39** [1], 177-183

Julkunen, P., Wilson, W., Jurvelin, J. S., Rieppo, J., Qu, C. J., Lammi, M. J., and Korhonen, R. K. 2008. Stress-relaxation of human patellar articular cartilage in unconfined compression: Prediction of mechanical response by tissue composition and structure. *Journal of Biomechanics* **41** [9], 1978-1986

Klein, T. J., Schumacher, B. L., Schmidt, T. A., Li, K. W., Voegtline, M. S., Masuda, K., Thonar, E. J. M., and Sah, R. L. 2003. Tissue engineering of stratified articular cartilage from chondrocyte subpopulations. *Osteoarthritis and Cartilage* **11**, 595-602

Laasanen, M. S., Saarakkala, S., Toyras, J., Rieppo, J., and Jurvelin, J. S. 2005. Site-specific ultrasound reflection properties and superficial collagen content of bovine knee articular cartilage. *Physics in Medicine and Biology* **50** [14], 3221-3233

Ma, P. X., Schloo, B., Mooney, D., and Langer, R. 1995. Development of Biomechanical Properties and Morphogenesis of In-Vitro Tissue Engineered Cartilage. *Journal of Biomedical Materials Research* **29** [12], 1587-1595

Mak, A. F. 1986. Apparent viscoelastic behavior of articular cartilage - The contributions from the intrinsic matrix viscoelasticity. 2. *Journal of Biomechanical Engineering-Transactions of the Asme* **108**, 123-130

Mansour, J. M. and Mow, V. C. 1976. Permeability of Articular-Cartilage Under Compressive Strain and at High-Pressures. *Journal of Bone and Joint Surgery-American Volume* **58** [4], 509-516

Maroudas, A. 1979. Adult Articular Cartilage. Freeman, M. A. R. Adult Articular Cartilage. 2, 215-290 Tunbridge Wells, England, Pitman Medical.

Mow, V. and Guo, X. E. 2002. Mechano-electrochemical properties of articular cartilage: Their inhomogeneities and anisotropies. *Annual Review of Biomedical Engineering* **4**, 175-209

Mow, V. C., Zhu, W., and Ratcliffe, A. 1991. Structure and function of articular cartilage and meniscus. Mow, V. C. and Hayes, W. C. Basic Orthopaedic Biomechanics. 143-198 New York, Raven Press.

National Instruments Corporation. 2008. LabVIEW. Austin, Texas, USA. 2001.

Nieminen, H. J., Saarakkala, S., Laasanen, M. S., Hirvonen, J., Jurvelin, J. S., and Toyras, J. 2004. Ultrasound attenuation in normal and spontaneously degenerated articular cartilage. *Ultrasound in Medicine and Biology* **30** [4], 493-500

Ophir, J., Cespedes, I., Ponnekanti, H., Yazdi, Y., and Li, X. 1991. Elastography - A Quantitative Method for Imaging the Elasticity of Biological Tissues. *Ultrasonic Imaging* **13** [2], 111-134

Park, S., Nicoll, S. B., Mauck, R. L., and Ateshian, G. A. 2008. Cartilage mechanical response under dynamic compression at physiological stress levels following collagenase digestion. *Annals of Biomedical Engineering* **36** [3], 425-434

Patil, S. G., Zheng, Y. P., Wu, J. Y., and Shi, J. 2004. Measurement of depth-dependence and anisotropy of ultrasound speed of bovine articular cartilage in vitro. *Ultrasound in Medicine and Biology* **30** [7], 953-963

Righetti, R., Ophir, J., and Krouskop, T. A. 2005. A method for generating permeability elastograms and Poisson's ratio time-constant elastograms. *Ultrasound in Medicine and Biology* **31** [6], 803-816

Soulhat, J., Buschmann, M. D., and Shirazi-Adl, A. 1999. A fibril-network-reinforced biphasic model of cartilage in unconfined compression. *Journal of Biomechanical Engineering-Transactions of the Asme* **121** [3], 340-347

The MathWorks, Inc. 2008. Matlab. 6.1 Natick, Massachusetts, USA. 2004.

Toyras, J., Rieppo, J., Nieminen, M. T., Helminen, H. J., and Jurvelin, J. S. 1999. Characterization of enzymatically induced degradation of articular cartilage using high frequency ultrasound. *Physics in Medicine and Biology* **44** [11], 2723-2733

Verteramo, A. and Seedhom, B. B. 2007. Effect of a single impact loading on the structure and mechanical properties of articular cartilage. *Journal of Biomechanics* **40** [16], 3580-3589

Yu, H. S., Grynpas, M., and Kandel, R. A. 1997. Composition of cartilagenous tissue with mineralized and non-mineralized zones formed in vitro. *Biomaterials* **18** [21], 1425-1431

Zheng, Y. P., Mak, A. F. T., Lau, K. P., and Qin, L. 2002. An ultrasonic measurement for in vitro depth-dependent equilibrium strains of articular cartilage in compression. *Physics in Medicine and Biology* **47** [17], 3165-3180

Zheng, Y. P., Niu, H. J., Mak, F. T. A., and Huang, Y. P. 2005. Ultrasonic measurement of depth-dependent transient behaviors of articular cartilage under compression. *Journal of Biomechanics* **38** [9], 1830-1837

Journal of Physics: Conference Series **195** (2009) 012005 doi:10.1088/1742-6596/195/1/012005

SU-8 photoresist and SU-8 based nanocomposites for broadband acoustical matching at 1 GHz

A. Ndieguene[1, 2, 3], P. Campistron[1, 2, 3], J. Carlier[1, 2, 3], S. Wang[1, 2, 3], D. Callens-Debavelaere[1, 2, 3], B. Nongaillard[1, 2, 3]

E-mail: Assane.Ndieguene@meletu.univ-valenciennes.fr

[1] Univ Lille Nord de France, F-59000 Lille, France
[2] UVHC, IEMN-DOAE, F-59313 Valenciennes, France
[3] CNRS, UMR 8520, F-59650 Villeneuve d'Ascq, France

Abstract. So as to integrate acoustic functions in BioMEMS using 1 GHz ZnO transducers deposited on silicon substrates, acoustic waves propagation through the silicon substrate and its transmission in water needs to be maximized (the insertion losses at the Si / water interface are about $6dB$).

In the context of integration, it is interesting for mechanical impedance matching to use photosensitive materials such as SU-8 so that patterns may be obtained. Nanocomposite materials based on SU-8 mixed with nanoparticles having adequate impedances were fabricated. These new materials are characterized in terms of their acoustic velocity, impedance and attenuation. For this, the nanocomposite layers are deposited on the substrate by spin coating to obtain a thickness of about 10 μm, in order to separate acoustic echoes from the material (even if $\lambda/4$ layer thickness is lower than 1μm).

The insertion losses of the device immersed in water can be simulated as a function of frequency for a given reflection coefficient between the silicon substrate and the photoresist. The characteristics of some nanocomposites made with SU-8 and various concentrations of nanoparticles like TiO_2 , $SrTiO_3$ or W have been determined.

1. Introduction

Recent advances in biology, and more specifically biological detection, have revealed the need to work with higher resolutions on a micrometer scale. This implies an increase in the frequency range used, that is to say in the order of gigahertz [1-2]. This frequency range can be obtained using a ZnO piezo-electric transducer. However, as it can be considered that the cell's environment is mainly water, much of the energy of the acoustic waves emitted is reflected at the Si / water interface due to the huge impedance mismatch between these two materials. In addition, the attenuation in water is high (about 220dB/mm/ GHz^2); the acoustic waves will thus have difficulty to reach the cells.

The aim of our study is to optimize the energy transfer between the silicon substrate and water, which implies good mechanical matching between these two materials. The standard solution is to match the impedance using one or more quarter wavelength transmission lines with specific acoustic impedance. It is well known that the more segments, the higher the bandwidth. As materials having the required impedance are generally not available we must create them [3 for example] and doing that, try also to minimize their absorption. The way we chose is to mix nanosized particles in SU-8. SU-8 is an epoxy-based negative photoresist that can be patterned for high ratio structures [4]. So it makes it possible to use it for Lab-On-Chip fabrication [5-6].

Theory of multi section quarter wavelength transformer [7] is used to compute the required values of acoustical impedance optimizing the matching of water to silicon. Mechanical impedances to use are about 5.5 MRayls for one section quarter wavelength transformer, 8.8 MRayls and 3.4 MRayls for a two section quarter wavelength transformer, 12.56 MRayls, 5.48 MRayls and 2.39 MRayls for a three section quarter wavelength transformer, and so on. The two sections quarter wavelength transformer can be easily realized by glass and polymers deposition.

In addition to pure SU-8, W, TiO_2 and $SrTiO_3$ nanoparticles are used to fabricate different composites with adjustable acoustic impedances. The fabrication process and characterization of high frequency ultrasound propagation in these materials are presented in this paper. We check that the photoresist property of the SU-8 resin remain in the low concentration mixture we use. The phase velocity and the attenuation coefficients of the acoustic waves are also measured.

2. Material fabrication and characterization

2.1. Matching layer fabrication

The fabrication of the 1GHz transducers has been presented previously [8]. The SU-8 2000 photoresist and TiO_2 nanosize particles for example (TiO_2: 35 nm diameter, *Degussa Corporation*; $SrTiO_3$: 50nm diameter, *Advanced Materials*) were ball-milled (*Retsch* PM100 planetary ball miller) in an agate jar in order to obtain an homogeneous nanocomposite. SU-8 2000 thinner (cyclopentanone, *Microchem Corporation*) was used as a solvent during ball milling. The mixture was deposited on the back of the silicon substrate by spin coating to obtain a thickness of about 10 μm. The substrate was then placed on a hot plate to evaporate the solvent. The thickness and surface roughness of the matching layer was characterized as less than 100 nm using a *Tencor Alpha step* profiler. Figure 1 shows a Scanning Electron Microscopy (SEM, *Zeiss* U55) surface image of a TiO_2/SU-8 nanocomposite and a cross section image of a W/SU-8 nanocomposite. It was revealed that the nanoparticles were homogenously distributed in the composite.

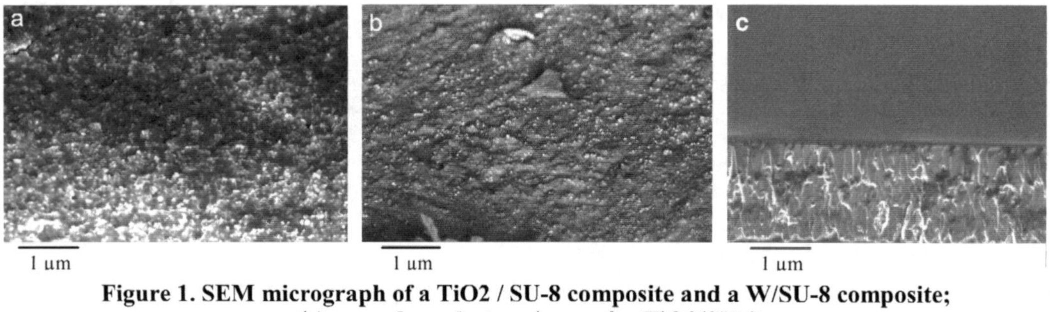

Figure 1. SEM micrograph of a TiO2 / SU-8 composite and a W/SU-8 composite;
(a) secondary electron image for TiO2/SU-8
(b) backscattered electron image for TiO2/SU-8
(c) Cross section of W/SU-8 nanocomposite.

2.2. Lithography of the nanocomposites

The lithographic efficiency of SU-8-based nanocomposites was studied. The lithography process is as follows: spin-coating (*Suss Microtech Spin Coater*, speed: 3000 rpm; time: 20 seconds) of the nanocomposite layer, then pre-baking at 95 °C for 4 minutes. The sample is then exposed to 365 nm UV for 20 seconds at 13 mW/cm^2, post-baking at 95 °C for 4 minutes. The development is made in SU-8 developer for 30 seconds. Isopropyl alcohol (IPA) was used to rinse the developed samples. Figure 2 shows the photo-defined pattern of an SU-8-based nanocomposite measured by FIB (*FEI Strata* Focus Ion Beam). As a consequence, the nanocomposites can be used in micro fabrications as SU-8 and also for the integration of acoustical characterization in lab-on-chip technology. These results show the possibility to use the nanocomposites also for high frequency piezocomposite for which it is necessary to adjust the mechanical impedance of the non piezoelectric material.

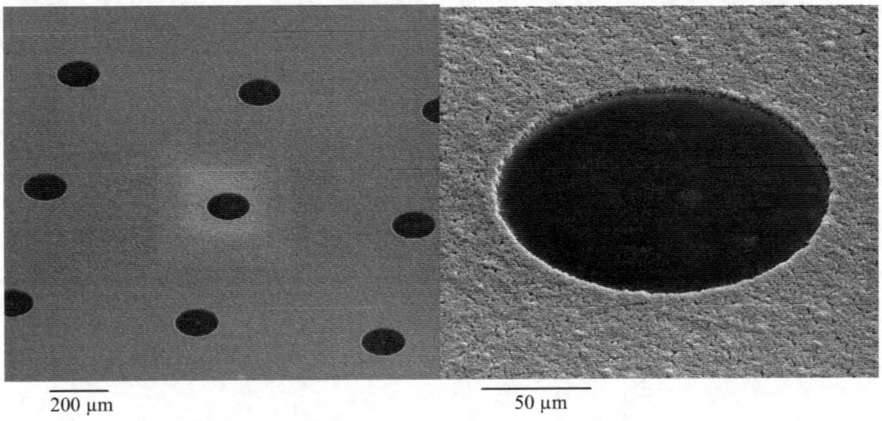

Figure 2. FIB images of the lithography of an SU-8-based nanocomposite. The holes patterned in the composite has a diameter of 150 µm

2.3. Determination of acoustic properties

To measure the acoustic properties of the materials, a broadband (300 kHz-8 GHz) vector network analyzer (*Rohde Schwarz,* ZVA 8) was used with a prober system (*Suss Microtech* EP4), as shown in figure 3.

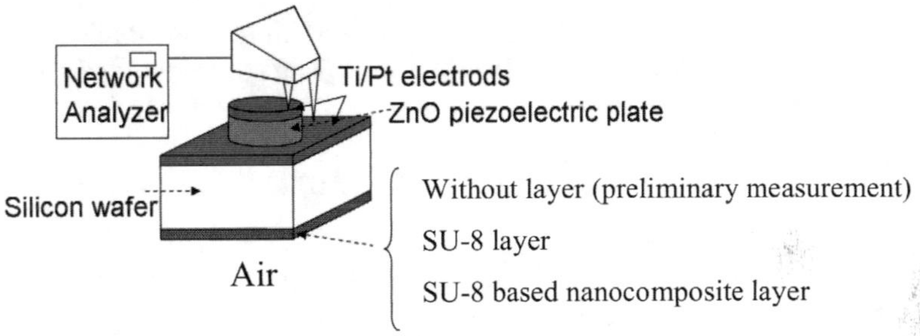

Figure 3. Scheme of the device under test
The S_{11} parameter measurement is obtained thanks to a prober system connected to a VNA

This measurement uses the parameter S_{11}, which is the ratio between the reflected and the incident electrical wave. In figure 4, the real and imaginary parts of S_{11} are plotted versus the frequency. This signal is extracted directly from the vector network analyzer. The real and imaginary parts of S_{11} show oscillations superimposed on slow variations. Oscillations are produced by the propagation of acoustic waves reflected on both sides of the silicon wafer and slow variations from the electrical behavior of the transducer. It has already been shown that [9]:

$$S_{11}(f) = S_{11}^{el}(f) + K\, S_{11}^{ac}(f) \tag{2}$$

Where:

K is the electro-acousto-electric conversion coefficient of the transducer.

S_{11}^{ac} is the sum of the acoustic waves that are formed from all the possible reflections before returning to the piezo-electric plate. This parameter can be used to estimate the acoustic characteristics of the materials.

S_{11}^{el} is the electrical reflection on the transducer. It allows the electrical impedance of the transducer to be measured. A Tchebychev type II high pass filter is applied in order to isolate the acoustical response (oscillations) from the electrical response (slow variations).

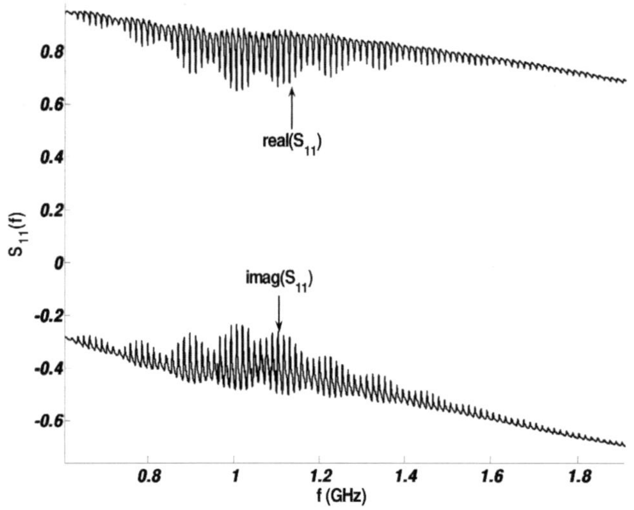

Figure 4. Real and imaginary parts of the measured S_{11} parameter versus frequency for the preliminary test (Transducer / Silicon / Air).

The impulse response of the system $S_{11}(t)$ is obtained by computing the inverse Fourier transform [10] of the $S_{11}(f)$ parameter. It is formed by a set of peaks centered at the delay time of the arrival of waves. A first measurement without layer is needed to characterize the electro-acousto-electric conversion coefficient of the transducer. Then an other measurement of the S_{11} parameter of the system for which the thin layer under test has been spin coated is performed. The comparison of the first echoes allows to evaluate reflection coefficient at the interface silicon / layer under test, the other echoes allow to evaluate the attenuation coefficient and the velocity of the ultrasonic wave propagating in the layer under test.

Figure 5 shows the impulse responses of signals associated with the silicon / air (preliminary test) and the silicon / SU-8 / air configurations. The comparison of the two curves shows that the first impulse decreases when the reflection coefficient at the Si/SU-8 interface is lower than that at the Si/air interface. It is obvious as a part of the acoustic energy is transmitted to the layer. Other impulses come from the reflection of ultrasonic waves on both of the SU-8 layer. As they are delayed and absorbed by each travel in the sample, they can be used as follows to characterize the different materials.

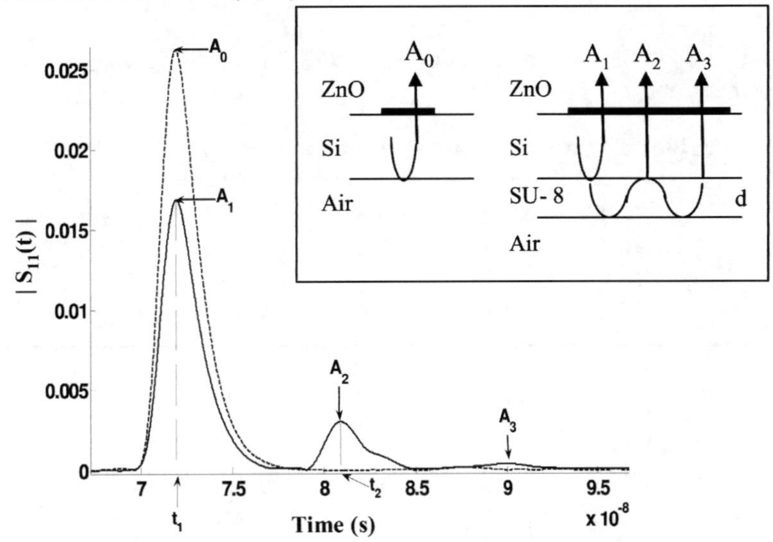

Figure 5. Impulse responses computed from S_{11} parameter versus time.
Dashed line : without layer (preliminary test)
Solid line : about 10 µm thin SU-8 layer
Insert : waves travels in the systems

The phase velocity in the material is given by:

$$V_a = \frac{2*d}{t_2 - t_1} \qquad \textbf{(3)}$$

To calculate the impedance and attenuation, we proceed as follows:

The reflection coefficient of the Si / material interface is given by: $r = \frac{A_1}{A_0}$,

Amplitudes A_0 and A_1 correspond to the maximum amplitude of the reflected acoustic signal at the Si / Air interface and the maximum amplitude of the reflected acoustic signal at the Si / Material interface respectively. A_2 and A_3 correspond to the maximum amplitude of the reflected acoustic signal at the Si / material interface after respectively 2 or 3 reflections in the silicon.

The reflection coefficient r is also defined by: $r = \frac{Z_a - Z_{Si}}{Z_a + Z_{Si}}$ **(4)**

This thus gives the material's acoustic impedance:

$$Z_a = \frac{1+r}{1-r} * Z_{Si} \qquad \textbf{(5)}$$

The attenuation in the material can be estimated by:

$$\alpha = -\frac{1}{2*d} * \log\left(\frac{A_3}{r*A_2}\right) \qquad \textbf{(6)}$$

We use the same method to characterize all the other samples.

3. Results and Discussion

3.1. Acoustical characteristics of the materials

Nanocomposites made from SU-8 and W (or TiO_2 or $SrTiO_3$) nanoparticles have been characterized. The results are presented in Table 1. The concentration of SU-8 in the nanocomposite as shown in table

1 is the concentration of SU-8 with its thinner (SU-8 2000 thinner evaporates during ball milling) and not its "dry" concentration.

Table 1. Acoustic characteristics of nanocomposites at 1 GHz

Acoustic characteristics	SU-8	SU-8 90% TiO_2 10%	SU-8 80% TiO_2 20%	SU-8 95% W 5%	SU-8 80% $SrTiO_3$ 20%	SU-8 70% TiO_2 30%	SU-8 70% TiO_2 15% $SrTiO_3$ 15%	SU-8 70% $SrTiO_3$ 30%
Impedance (MRayls)	3.5	4.5	5	5.3	5.5	6	6.5	7
Phase velocity (m/s)	2890	2490	2710	2700	2640	2600	2500	2370
Attenuation (dB/µm)	0.3	0.5	0.6	0.4	0.3	0.5	0.6	0.7

The impedance of pure SU-8 is lower than the impedance of SU-8 mixed with a nanocomposite. In the case of a low concentration mixture, the phase velocity remains almost constant, so the impedance is governed by the increase in density. It was found that the concentration of particles in the mixture could not exceed 30% for TiO_2 and $SrTiO_3$ and 15% for W as the solution becomes inhomogeneous and therefore inoperable. So, an acoustic impedance higher than 7 MRayls could not be obtained and this technique is limited to a one section quarter wavelength transformer. We will therefore compare the efficiency of the impedance matching of one nanocomposite layer with the two section quarter wavelength SiO_2/SU-8/H2O commonly used and characterized in a previous study [8].

3.2. Acoustical matching

After obtaining the experimental results of the acoustic characteristics of the materials, the insertion loss at the interface between the different quarter wavelength matching layers and water as a function of frequency were determined as shown in figure 6. The insertion loss at the interface between silicon and water is close to 6 dB. When SU-8-based nanocomposites were introduced as matching layers, the insertion loss greatly decreased. The insertion loss is given by:

$$L = 10\log_{10}\left[1-\left|r(f)\right|^2\right] \quad \textbf{(7)}$$

r(f) is computed by filtering the oscillation of S_{11}(f) so as to conserve only these corresponding to the first arrival of the impulse response [8].

In figure 6, an example of two section acoustical layer matching is presented. A quarter wavelength SU-8 layer and a quarter wavelength SiO_2 layer are stacked. The experimental and theoretical results, based on transfer matrix calculation were compared. A simulation with only one nanocomposite layer is also presented. The bandwidth obtained is greatly improved compared to a single quarter wavelength layer.

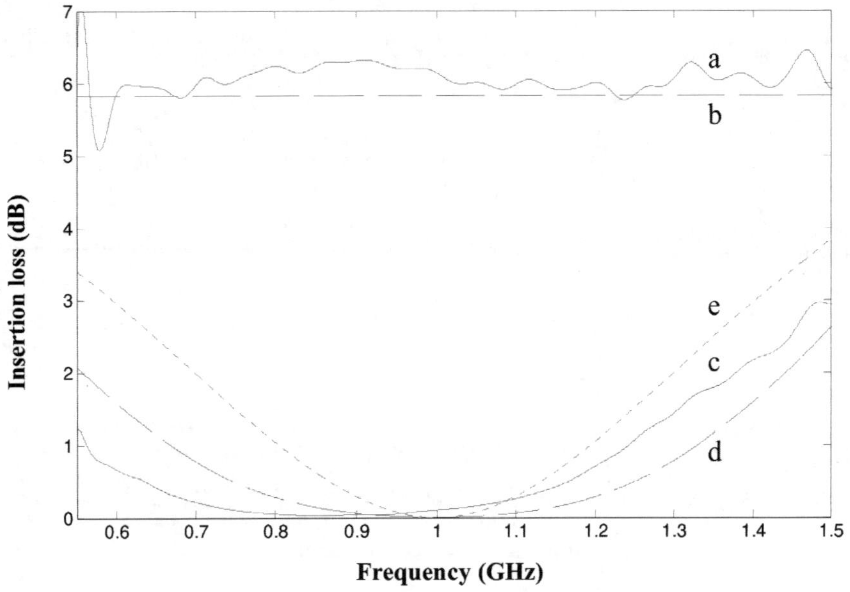

**Figure 6. Insertion loss (dB) versus frequency (GHz) for:
the Si/water interface ((a) measured, (b) theoretical)
the Si/SU-8/SiO_2/water interface ((c) measured, (d) theoretical)
the Si/SU-8 $_{80\%}$ $SrTiO_3$ $_{20\%}$/water interface (e) theoretical)**

The minimum is centered around 0.9 GHz. This difference can be explained by an imprecision in the thickness the spin coating of the layer and by variation in the phase velocity in the layer, which is strongly dependent on the fabrication process.

4. Conclusion

This study has shown that new materials could be created. These materials have been characterized and we have presented their acoustic properties (velocity, attenuation coefficient and acoustic impedance). The acoustic impedance can be easily adjusted between 3 MRayls and 7 MRayls. The obtained materials remain photosensitive and so, can be patterned and have potential applications for the integration of acoustical characterization in lab-on-chip technology. The range of acoustic impedance we can get is not high enough for making a two section impedance transformer using only polymers. So we realize and characterize the two sections efficient impedance matching (Si / SiO2 / SU-8 / H2O) as shown on curves of the figure 6. The attenuations of ultrasonic 1GHz longitudinal waves in these new materials remain weak. So, used as matching layer, they can increase energy transmitted from silicon to water of about 5dB.

5. Acknowledgments

We would like to acknowledge Pr Remiens' team and Pr Descamps' team and all the IEMN staff for their kind help in the technological steps and measurements.

6. References

[1] T. Kundu, J. Bereiter-Hahn, and I. Karl, "Cell Property Determination from the Acoustic Microscope Generated, Voltage Versus Frequency Curves", Biophysical Journal, vol. 78, pp 2270-2279, 2000.

[2] T. Kundu, J. -P. Lee, C. Blase and J. Bereiter-Hahn, "Acoustic microscope lens modeling and its application in determining biological cell properties,from single- and multi-layered cell models," J. Acoust. Soc. Am., vol. 120, no. 3, pp 1646-1654, 2006.

[3] M.I.Haller, B.T.Khuri-Yakub, "Tapered acoustic matching layers", Ultrasonic symposium, pp505-508, 1993.

[4] Lorenz H, Despont M, Fahrni N, Lablance N, Renaud P and Vettiger P "a low cost negative resist for MEMS", J. Micromech. Microeng. 7, pp 121–124, 1997.

[5] J. Carlier, S. Arscott , V. Thomy, J.C. Fourrier, F. Caron, J. C. Camart , C. Druon, and P. Tabourier, "Integrated Microfluidics based on Multi-Layered SU-8 for Mass Spectrometry analysis", Journal of Micromechanics and Microengineering, 14, 4,pp 619-624, 2004.

[6] J. Carlier, K. Chuda, S. Arscott, V. Thomy, B. Verbeke, X. Coqueret, J.-C. Camart, C. Druon and P. Tabourier, "High pressure-resistant SU-8 microchannels for monolithic porous structure integration", Journal of Micromechanics and Microengineering. 16, pp 2211-2219, 2006.

[7] Mathaei, Young, Jones Microwave Filters, Impedance-Matching Networks, and Coupling Structures McGraw-Hill, 1964.

[8] S. X. Wang, P. Campistron, J. Carlier, D. Callens-Debavelaere, B. Nongaillard, A. NDieguene, G. Nassar, C. Soyer, "Acoustical properties characterization of a composite made of SU8 and nanoparticles for BioMEMS application," Acoustic' 08, Paris, May 2008.

[9] Y. Deblock, P. Campistron, M. Lippert, and C. Bruneel, "Electrical characterization of plate piezoelectric transducers bonded to a finite substrate," J. Acoust .Soc .Am. vol. 111, no. 6, pp. 2681-2685, 2002.

[10] Y. Deblock, P. Campistron, and B. Nongaillard, "A continuous wave method for ultrasonic characterization of liquids materials," J. Acoust. Soc. Am. vol. 118, no. 3, pp. 1388-1393, 2005.

Strong nonlinear harmonic generation in a PZT/Aluminum resonator

D Parenthoine[1,2], L Haumesser[1], F Vander Meulen [1] and L-P Tran-Huu-Hue[1,2]

[1] University Francois Rabelais of Tours, U 930 Imagerie et Cerveau, CNRS 2448, ENIVL, rue de la Chocolaterie, BP 3410, 41034 Blois, France

[2] ENIVL, rue de la Chocolaterie, BP 3410, 41034 Blois, France

E-mail: parenthoine@univ-tours.fr

Abstract. In this work, the extentional vibration mode of a coupled PZT/ Aluminum rod resonator is studied experimentally. Geometrical characteristics of the PZT are its 27 mm length and its 4x4 mm^2 cross section area. The excitation voltage consists in sinusoidal bursts in the frequency range (20-80 kHz). Velocity measurements are performed at both ends of this system, using a laser probe. Strong harmonic distortions in the mechanical response (up to -20 dB with respect to the primary wave amplitude) have been observed. The corresponding input levels are far lower than those which are necessary to observe quadratic second harmonic generation in a free PZT resonator. The strong nonlinear effect can be explained as a super-harmonic resonance of the system due to a specific ratio between the eigen frequencies of the two parts of the resonator. Evolution of fundamental and harmonic responses are observed as a function of input levels, highlighting hysteretic behavior.

1. Introduction

It is well-known that the addition of a load on a resonator leads to a coupled system whose vibratory behavior can be completely different from the one of the free resonator [1]. For a linear system, the result is the modification of eigen frequencies. When the system is nonlinear, it induces change in the harmonic content of the oscillator. In previous works [2], [3], a quadratic regime of nonlinearity has been highlighted on a soft PZT piezoelectric resonator. The results have then allowed third-order constants to be indentified [3]. In the present study, we used an aluminum load - this material can be considered as quite linear compared to soft PZT - with respect to the PZT length to modify the linear resonances, then the nonlinear behavior. The results in terms of harmonic generation are analysed.

2. Experimental protocol

Figure 1 shows the experimental set-up. Voltage bursts, delivered by a frequency generator and amplified, are applied to a piezoelectric rod 27 x 4 x 4 mm^3 in series with a 4.7 $k\Omega$ resistance. A 500-cycles sine wave has been chosen to reach a steady state. Moreover, repetition rates were low enough to avoid any heating effects that could influence measurements. The electrodes are placed at each edge of the lateral surface in order to reduce the perturbation of the mechanical displacement. The load is an aluminum cylinder rod with a 35 *mm* length and a 5 *mm* diameter. It is fixed at one end of the piezoelectric sample using phenyl salicylate salt (salol). The coupled system is lying on a suppport in order to avoid any fixation which could disturb the extensionnal response of the system. Voltage measurements are performed using a digital oscilloscope and electrical admittance can be deduced from V_0 and V_r (Fig 1). A laser vibrometer is used to measure the velocity at each end of the system, successively.

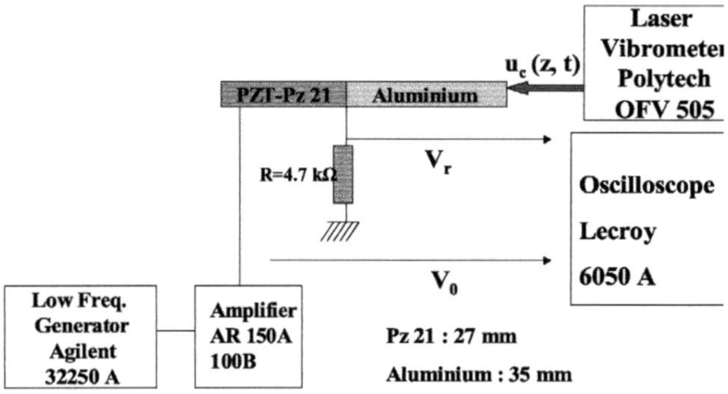

Figure 1. Experimental set-up.

3. Measurements

3.1. Linear measurements

Preliminary measurements at low level of excitation (1 V/cm) have been performed in order to determine the linear response of the structure (Fig 2). Thus, resonance peaks were identified at respectively 27.5 kHz and 60 kHz. One can see that the maxima of electrical admittance

(Fig 2.b) correspond to the maxima of mechanical response (Fig 2.a) which confirms the one dimensional behavior in length-extension mode. The ratio of the two first eigen frequencies, approximately 2, has been chosen.

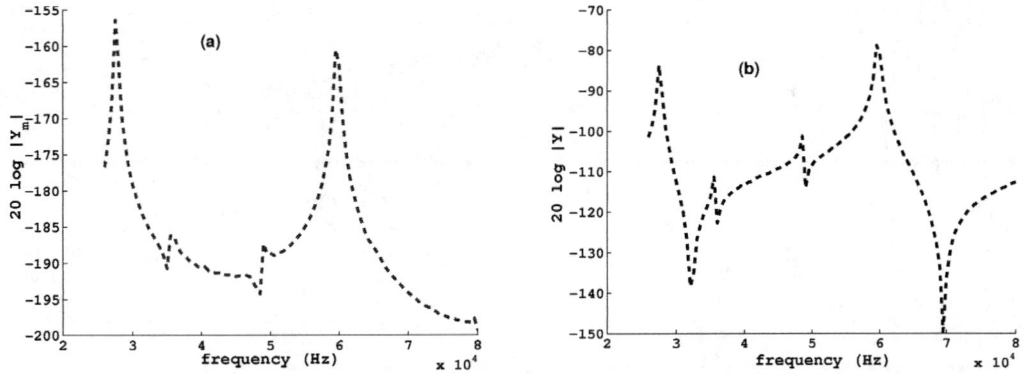

Figure 2. Linear response of the system: (a) Modulus of the electromechanical transfer function at the aluminum free end and (b) modulus of the electrical admittance versus frequency. Vertical scales are logarithmic: a dB scale is used where the reference is respectively 1 m/s and 1 S.

3.2. Nonlinear measurements

Measurements at increasing levels were then performed in order to observe nonlinear effects for different frequencies of excitation near the first and second resonance peaks. The comparison between applied voltages and velocity responses in terms of time signals and corresponding spectra is given below (Fig 3). Distortion levels reaching -40 dB are measured in the electrical voltage source. In the corresponding response, the measured distortion levels, reach -15 dB for second and third harmonic components. Such an increase of the distortion of the velocity spectra, cannot be due to the voltage source if we consider the electromechanical transfer function close to 55 kHz corresponding to 2 times the 27.5 kHz excitation frequency(Fig 2). Then, such levels can only be due to internal nonlinearity of the resonator.

3.3. Nonlinear regime

Second harmonic generation has been then studied as a function of the square of fundamental amplitudes for the following frequencies of excitation: 27.5 kHz (corresponding to the first resonance peak), 29 kHz just after the peak and 60 kHz (corresponding to the second resonance peak). From results plotted in Fig 4, one can see that for the excitation at 60 kHz only, the regime of second harmonic generation remains quadratic in the range of used levels of excitation (up to 50 V/cm). This case is quite comparable to phenomena previously observed on a free PZT-resonator [2][3]. On the contrary, the results for 27.5 kHz and 29 kHz clearly show that the increase of harmonic generation is higher than quadratic even for low levels of excitation. In the case of an excitation close to the first resonance peak, the fundamental response reaches a limit whereas the second harmonic is still increasing. This limit is not observed for other frequencies of excitation such as 60 kHz. This phenomenon appears as a transfer of energy towards higher frequencies. These high levels of harmonic generation while exciting at the first resonance peak can be explained by the geometry of the system where the first and second eigen modes are tuned in a ratio of two. Such phenomenon can be considered as a super-harmonic resonance for

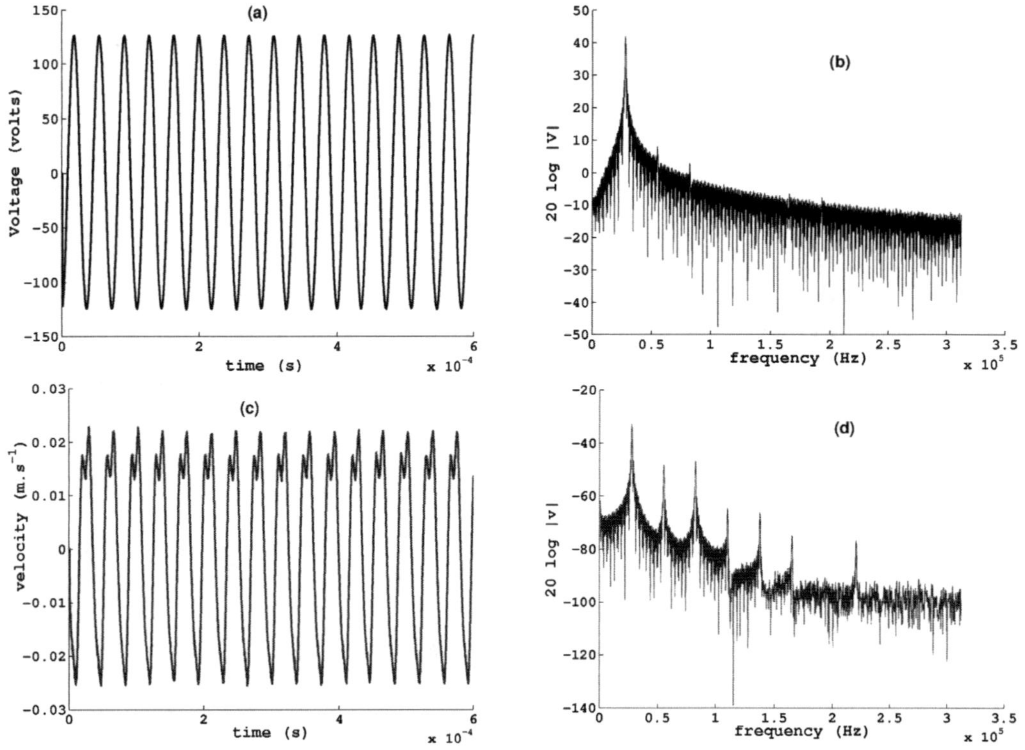

Figure 3. (a) Signal of the highest electric excitation used at 27.5 kHz, (b) spectra of the previous signal, (c) and (d) corresponding velocity responses.

the nonlinear system [4].
Results presented in Fig 4 have been obtained using increasing voltages. In addition, the observation using increasing and decreasing voltages has highlighted hysteresis phenomenon (Fig 5). One can see that the hysteresis loop is more pronounced close to the first resonance peak. Such hysteretic phenomena have also been observed on a free resonator [5] but for higher levels of excitation (100 V-peak /cm). However, using these levels, no subharmonic component which can be observed for more specific geometries [6] has appeared in our configuration.

4. Conclusion

The present study has highligthed the possibility to obtain higher regime of second harmonic generation in a PZT/Aluminum coupled resonator compared to the one of the free PZT resonator. Thus, an increase of the distortion level of the mechanical response to a single frequency of excitation appears when the eigen frequencies are tuned at a ratio of two. In addition, a saturation of the fundamental component and an hysteretic behavior are then observed. The result may have consequences for different applications of PZT materials such as those who require nonlinearity level control, namely harmonic imaging for medical diagnostics or nonlinear NDE. Other applications in which very high power ultrasonics are generated, such as long distance SONAR, industrial cleaning and welding, sonochemistry or medical therapy, may also benefit from these observations and their analysis.

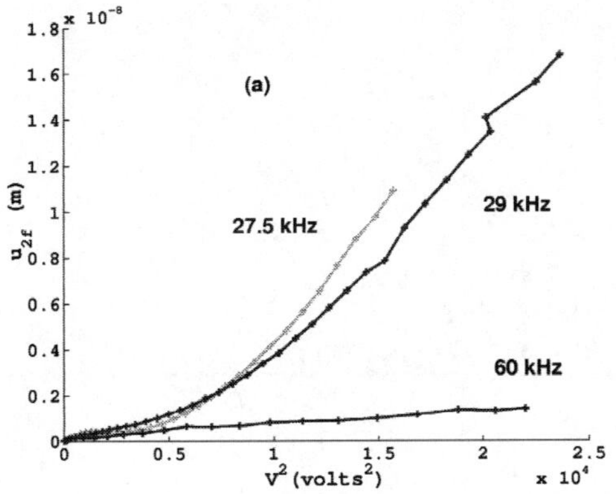

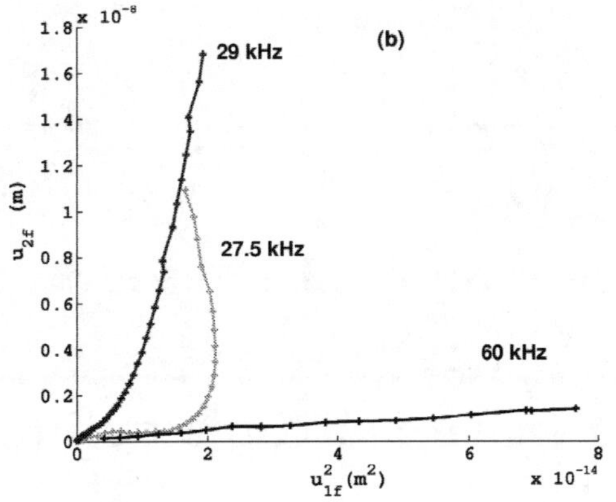

Figure 4. Evolution of the second harmonic component in displacement versus: (a) the square of the applied voltage and (b) the square of the fundamental component.

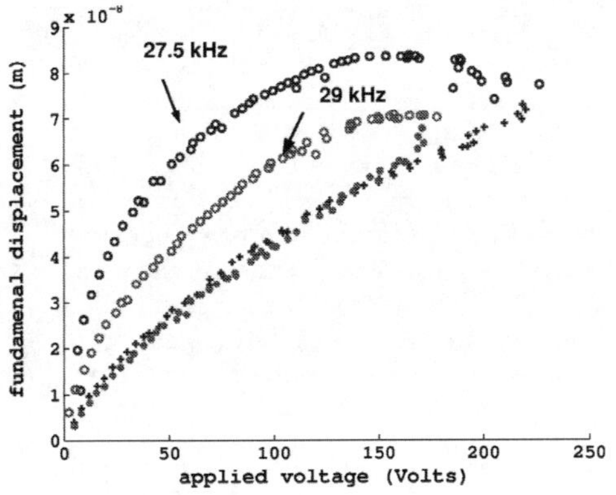

Figure 5. Hysteretic phenomenon in the mechanical response at 27.5 kHz and 29 kHz. Circles: response obtained by increasing voltages, crosses: response obtained by decreasing voltages.

5. References

[1] E. Dieulesaint and D. Royer *Acoustique du solide, tome 2*, Masson, 1997.

[2] W. Jiang, S. Zhou, X. Li and Y. Shui, "Second harmonic generation in piezoelectric resonators", *IEEE, Ultrasonics Symposium*, pp 969-972, 1994.

[3] D. Parenthoine, L. Haumesser, F. Vander Meulen, M. Lethiecq and L.-P. Tran-Huu-Hue " Nonlinear constant evaluation in a piezoelectric rod from analysis of second harmonic generation " *IEEE Trans. Ultrason. Ferroelec. Freq. Contr*, vol 56, n0 1, pp 167-174, 2009.

[4] A. H. Nayfeh and D. T.Mook, *Nonlinear oscillations*, Willey, New York, 1979.

[5] J. Blackburn and M. Cain, "Nonlinear piezoelectric resonance: a theoritically rigorous approach to constant i/v measuements", *J. Appl. Phys.*, 114101, vol 100, 2006.

[6] A. Bettucci, A. Biagoni, A. D'Orazio and D. Paseri, " Periodic oscillations within the chaotic region in finite piezoelectric structures" Proc in the 18th Int. Symp. on Nonlinear Acoustics, pp 251-254, 2008.

Semi-analytical model of acoustic-wave generation by a laser line pulse in an optically absorptive isotropic cylinder

D Ségur[1], A L Shuvalov[1], Y Pan[2], B Audoin[1]

[1] Laboratoire de Mécanique Physique, UMR CNRS 8469, Université Bordeaux 1, 33405 Talence, France

[2] Institute of Acoustics, Tongji University, Shanghai, China

E-mail: d.segur@lmp.u-bordeaux1.fr

Abstract. Semi-analytical model for calculating acoustic response to a line-focused laser pulse in an optically absorptive isotropic cylinder is proposed and implemented. It is assumed that the laser input is absorbed over the volume and thus creates a radially distributed thermoelastic source. Closed-form solution is obtained in the Fourier domain. Two inverse Fourier transforms in frequency and circumferential wave number yield the sought waveforms of acoustic response in the time-space domain. Numerical simulation is compared to the calculation based on a surface dipole source. The two signals have essentially different shapes of the wave-arrival peaks when the optical penetration is large enough.

Keywords: laser ultrasonics, distributed thermoelastic source, acoustic responce, cylinder

1. Introduction

The laser ultrasonics technique is a unique tool for acoustic-wave generation in objects with a curved surface. Much work has been done by this means for generating surface and bulk acoustic waves in opaque cylindrical targets with a view of their non-destructive testing and evaluation, see recent publications [1–4] and the bibliography therein. There exists a well developed theory for the case when, at a given laser wavelength, the target (either cylindrical or planar) is opaque and hence the source is located at the surface. Given so, the ablation model applies for high enough local energy such that causes ablation at the laser-heated region of the surface, and the dipole (Scruby's) model applies for relatively low incoming energy [5]. In this latter, thermoelastic regime, the dilatation of an infinitely small volume adjacent to the surface produces a surface source equivalent to a force dipole. The dipole strength has been derived in [6, 7].

Thermoelastic regime for the optically absorptive materials implies that the laser input penetrates into a target and creates a volume distribution of sources therein. Hence the assumption of a surface source is no longer appropriate. Laser generation of acoustic waves in optically absorptive materials has been considered for the case of a planar target (plate) [8]. The next natural step is extending laser ultrasonics treatment to optically absorptive cylindrical targets. To the best of the authors' knowledge, such a case has not yet been studied. The present paper proposes a two-dimensional (2D) semi-analytical model for calculating acoustic response to a line-focused laser pulse in an optically absorptive isotropic cylinder. Thermoelastic regime of generation is assumed and a typical case of materials with negligible thermal conductivity is treated. Closed-form solution is obtained in the Fourier domain to the inhomogeneous ordinary differential system with a radially distributed thermoelastic source and with the traction-free boundary condition at the cylinder surface. Performing two inverse Fourier transforms yields the sought waveforms of acoustic response in the time-space domain. This calculation is carried out in the paper for two different source settings. One of them formally treats the case of an opaque cylinder by assuming a vanishingly small optical penetration or, which is equivalent, a buried source tending to the cylinder surface. The other considers the actual case of an optically absorptive cylinder, where the exponentially decaying law of optical absorption produces a distributed source with large enough radial penetration length. The results obtained for these two settings are compared with the calculation based on Scruby's model of a surface dipole source. It is observed that the waveforms, computed for a buried source tending to the surface, agree very well with those following for Scruby's surface-source model, whereas the waveforms, computed for a distributed source, have essentially different shapes comparatively to the prediction following for the surface source. The former observation serves as a test for our model, while the latter highlights the effect of laser-energy penetration into the bulk of an optically absorptive cylinder.

2. Background

Consider an infinite homogenous isotropic elastic cylinder of radius R, whose surface is heated by the line-focused laser pulse independent of the axial coordinate z. The laser energy is supposed to be low enough to preclude ablation. It is assumed that viscoelasticity, thermal conduction and the contribution of deformation to the entropy change may be disregarded. The thermoelastic generation of acoustic wave is therefore described by the following simplified system of governing equations:

$$\nabla \cdot \sigma = \rho \ddot{\mathbf{u}}; \; \sigma = \lambda \mathrm{tr}\,(\varepsilon)\,\mathbf{I} + 2\,\mu\varepsilon - \beta T \mathbf{I}; \; \rho c_p \dot{T} = q \,, \tag{1}$$

where σ is the stress tensor including mechanical and thermal stresses, ε is the strain tensor, $\mathbf{u}$ is the displacement, ρ is the density, λ and μ are the Lamé coefficients, $\beta = (3\lambda + 2\mu)\,\alpha_T$ is the thermal modulus (α_T is the dilatation coefficient), T is the temperature rise, c_p is the specific heat, q is the power density of the line laser source, $\mathbf{I}$ is the identity matrix and tr denotes the trace.

The heat input $q(r, \theta, t)$ creates a volume distribution of thermal sources inside the cylinder through the optical absorption. Assuming normal incidence of the Gaussian laser beam [9] and the exponential fall off of the incoming energy E, it follows that

$$q(r,\theta,t) = \alpha\, E\, g(\theta)\, \delta(t)\, e^{-\alpha\,(R-r)}, \tag{2}$$

where

$$g(\theta) = \frac{1}{\gamma}\sqrt{\frac{4\ln 2}{\pi}}e^{-(4\,\ln 2\,)\frac{\theta^2}{\gamma^2}},\ \gamma = 2\ \mathrm{arctan}\,\frac{b}{2R} \tag{3}$$

describes the angular extent of the beam with width b. By $(1)_3$ and (2),

$$T(r,\theta,t) = \frac{\alpha E}{\rho c_p}\, g(\theta)\, H(t)\, e^{-\alpha(R-r)}, \tag{4}$$

where H is the Heaviside function. The temperature field (4) specifies the thermal stress $\sigma^{(T)} = -\beta T$ in (1).

Thus, the thermoelastic problem is reduced to solving the 2D plane-strain equation of motion for the displacement $\mathbf{u}(r,\theta,t) = u_r\mathbf{e}_r + u_\theta\mathbf{e}_\theta$ in the cylinder cross-section $(\mathbf{e}_r,\mathbf{e}_\theta)$. Divergence of the thermal stress plays the role of a source term. The equation is complemented by the traction-free boundary condition (BC) $\mathbf{e}_r\sigma = \mathbf{0}$ at the cylinder surface $r = R$.

3. Solution in the Fourier domain

3.1. *Setup*

Consider the above 2D problem in the Fourier domain $\{r,\nu,\omega\}$, where ν is the circumferential wave number and ω the angular frequency. Thereafter in this section the factor $e^{j(\nu\theta-\omega t)}$ is omitted, the same notations for the transformed functions are used, and their dependence on ν,ω is understood and suppressed. According to (1) and (4), we write

$$\begin{gathered}\sigma\,(r) = \sigma^{(m)}\,(r) + \sigma^{(T)}\,(r)\,\mathbf{I},\\ \sigma^{(m)}\,(r) = \lambda \mathrm{tr}\,[\varepsilon\,(r)]\,\mathbf{I} + 2\,\mu\varepsilon\,(r)\,,\ \ \sigma^{(T)}\,(r) = -\beta T_0 e^{-\alpha(R-r)},\end{gathered} \tag{5}$$

where $T_0\,(\sim\alpha)$ involves double transform of (4) in θ and t.

The equation of motion in the Fourier domain reduces to the ordinary differential system in $\mathbf{u}\,(r)$ with the source term provided by $\nabla\sigma^{(T)}\,(r)$. Using Helmholtz decomposition, this equation may be re-cast in the form of two uncoupled Bessel equations for the longitudinal and shear elastic potentials $\varphi\,(r)$ and $\psi\,(r)$,

$$\begin{pmatrix}\nabla^2 + k_L^2 & 0\\ 0 & \nabla^2 + k_T^2\end{pmatrix}\begin{pmatrix}\varphi\,(r)\\ \psi\,(r)\end{pmatrix} = \begin{pmatrix} be^{-\alpha(R-r)}\\ 0\end{pmatrix}, \tag{6}$$

where $k_L^2 = \rho\omega^2/\,(\lambda + 2\mu)\,,\ k_T^2 = \rho\omega^2/\mu$ and $b = \beta T_0/\,(\lambda + 2\mu)\,.$ The traction-free BC, which couples $\varphi\,(r)$ and $\psi\,(r)\,,$ is

$$\begin{pmatrix}\sigma_{rr}^{(m)}\,(R)\\ \sigma_{r\theta}^{(m)}\,(R)\end{pmatrix} + \begin{pmatrix}-\beta T_0\\ 0\end{pmatrix} = \mathbf{0}, \tag{7}$$

where

$$\begin{pmatrix} \sigma_{rr}^{(m)}(R) \\ \sigma_{r\theta}^{(m)}(R) \end{pmatrix} = \mathbf{S}(R)\begin{pmatrix} \varphi(R) \\ \psi(R) \end{pmatrix}, \ \mathbf{S}(r) = 2\mu \begin{pmatrix} \frac{\lambda}{2\mu}\nabla^2 + \frac{\mathrm{d}^2}{\mathrm{d}r^2} & \frac{j\nu}{r}\left(\frac{\mathrm{d}}{\mathrm{d}r} - \frac{1}{r}\right) \\ \frac{j\nu}{r}\left(\frac{\mathrm{d}}{\mathrm{d}r} - \frac{1}{r}\right) & \frac{1}{2}\nabla^2 - \frac{\mathrm{d}^2}{\mathrm{d}r^2} \end{pmatrix}. \tag{8}$$

For the future use, introduce the following abridged notations of the values taken at the boundary $r = R$:

$$K_{L,T} \equiv k_{L,T}R, \ B_{T,L} \equiv \frac{K_{T,L}J'_\nu(K_{T,L})}{J_\nu(K_{T,L})}, \ ' \equiv \left[\frac{\mathrm{d}J_\nu(x)}{\mathrm{d}x}\right]_{x=K_{L,T}}. \tag{9}$$

Note that expressing the results in terms of $B_{T,L}$ is helpful for numerical implementation, because the exponential growth of the Bessel functions of the order $\nu \gg K_{L,T}$ is avoided this way (see [10]).

3.2. Green function

First, we solve an auxiliary problem for a buried point source at r_0 ($0 \le r_0 < R$) under the homogeneous BC on the mechanical stresses (see $(8)_1$),

$$\begin{pmatrix} \nabla^2 + k_L^2 & 0 \\ 0 & \nabla^2 + k_T^2 \end{pmatrix}\begin{pmatrix} \Phi(r|r_0) \\ \Psi(r|r_0) \end{pmatrix} = \begin{pmatrix} bR^2\frac{\delta(r-r_0)}{r} \\ 0 \end{pmatrix}, \ \mathbf{S}(R)\begin{pmatrix} \Phi(R|r_0) \\ \Psi(R|r_0) \end{pmatrix}_{r_0<R} = \mathbf{0}, \tag{10}$$

where the factor R^2 in $(10)_1$ is added to merely keep physical dimensions intact. The partial solution to the given inhomogeneous Bessel equation on $\Phi(r|r_0)$ is [11]

$$g(r|r_0) = bR^2\frac{\pi}{2}\left[J_\nu(k_Lr)Y_\nu(k_Lr_0)H(r_0-r) + Y_\nu(k_Lr)J_\nu(k_Lr_0)H(r-r_0)\right], \tag{11}$$

where J_ν and Y_ν are the Bessel functions of the first and second kind. Thus the solution to (10) is sought in the form

$$\begin{pmatrix} \Phi(r|r_0) \\ \Psi(r|r_0) \end{pmatrix} = \begin{pmatrix} g(r|r_0) + C_\Phi(r_0)J_\nu(k_Lr) \\ C_\Psi(r_0)J_\nu(k_Tr) \end{pmatrix}, \tag{12}$$

where $J_\nu(k_Lr)$ and $J_\nu(k_Tr)$ are the solutions of the homogeneous Bessel equation on the potentials. Substituting (12) into the BC of (10) and appealing to $(8)_2$ provides an algebraic system of equations on the coefficients $C_\Phi(r_0)$ and $C_\Psi(r_0)$. It has the form

$$\mathbf{S}(R)\begin{pmatrix} \Phi(R|r_0) \\ \Psi(R|r_0) \end{pmatrix}_{r_0<R} = \frac{2\mu}{R^2}\left[\mathbf{\Lambda}\begin{pmatrix} C_\Phi(r_0) \\ C_\Psi(r_0) \end{pmatrix} + bR^2\frac{\pi}{2}J_\nu(k_Lr_0)\gamma\right] = \mathbf{0} \tag{13}$$

with

$$\mathbf{\Lambda} = \begin{pmatrix} \left(\nu^2 - \frac{K_T^2}{2} - B_L\right)J_\nu(K_L) & j\nu(B_T-1)J_\nu(K_T) \\ j\nu(B_L-1)J_\nu(K_L) & -\left(\nu^2 - \frac{K_T^2}{2} - B_T\right)J_\nu(K_T) \end{pmatrix},$$
$$\gamma = Y_\nu(K_L)\begin{pmatrix} \nu^2 - \frac{K_T^2}{2} - B_L^{(Y)} \\ j\nu\left(B_L^{(Y)} - 1\right) \end{pmatrix}, \tag{14}$$

where the use of (9) has been made, and $B_L^{(Y)}$ differs from B_L by replacing J_ν with Y_ν. Note that the matrix (first-rank tensor) $\mathbf{\Lambda}$ results in (13) from applying the boundary stress operator $\mathbf{S}(R)$ to the solutions of the homogeneous equation. Hence $D = 0$, where

$$D = \frac{\det \mathbf{\Lambda}}{J_\nu(K_L)\, J_\nu(K_T)} = -\nu^4 + \nu^2(B_L B_T + K_T^2 + 1) - \left(\frac{K_T^2}{2} + B_L\right)\left(\frac{K_T^2}{2} + B_T\right), \quad (15)$$

is the dispersion equation for the eigenmodes of the traction-free cylinder. By (13),

$$\begin{pmatrix} C_\Phi(r_0) \\ C_\Psi(r_0) \end{pmatrix} = -bR^2 \frac{\pi}{2} J_\nu(k_L r_0)\, \mathbf{\Lambda}^{-1}\gamma. \quad (16)$$

Inserting (16) along with (11) into (12) completes finding the solution $\Phi(r|r_0)$ and $\Psi(r|r_0)$ to (10).

3.3. Displacement for a distributed source

Once the solution $\Phi(r|r_0)$ and $\Psi(r|r_0)$ to (10) has been found, it remains to take its convolution with the radially distributed source $\sim e^{-\alpha(R-r)}$ for obtaining the solution $\varphi(r)$ and $\psi(r)$ to Eq. (6), and to invoke Helmholtz decomposition $\mathbf{u} = \nabla\varphi + \nabla \times (\psi \mathbf{e}_z)$ for finding $\mathbf{u}(r)$. Because these two are linear operations, they may be applied in the optional order. Thus

$$\mathbf{u}(r) = \frac{1}{R^2} \int_0^R \{\nabla\Phi(r|r_0) + \nabla \times [\Psi(r|r_0)\, \mathbf{e}_z]\}\, e^{-\alpha(R-r_0)} r_0 dr_0, \quad (17)$$

where $\mathbf{G}(r|r_0) = \nabla\Phi(r|r_0) + \nabla \times [\Psi(r|r_0)\, \mathbf{e}_z]$ is the displacement response to a buried source $\nabla\sigma^{(T)}(r) \sim \nabla\frac{\delta(r-r_0)}{r}$. Note that application of the boundary-stress differential operator to $g(r|r_0)$ (see (12)) in the integrand of (17) taken with $r, r_0 \to R$ provides an extra delta-function term, which compensates $\sigma^{(T)}(R) = -\beta T_0$ in the BC (7) for the distributed source.

In the context of laser ultrasonics, the important particular case is the displacement at the surface $r = R$. It can be written in the following closed form:

$$\mathbf{u}(R) = \begin{pmatrix} u_r \\ u_\theta \end{pmatrix} = IU_0 \frac{K_L^2}{J_\nu(K_L)} \begin{pmatrix} \nu^2 - B_T - \frac{K_T^2}{2} \\ j\nu(1 - B_T) \end{pmatrix}, \quad (18)$$

where

$$I = \int_0^1 J_\nu(K_L x)\, e^{-\alpha R(1-x)} x dx \;\; (x = r/R),\; U_0 = \frac{\beta T_0 R}{2\mu D}, \quad (19)$$

and D is given in (15).

Numerical simulations in §4.2 involve the model case of a buried source located at $r_0 = R - \epsilon$ tending to the surface. For this case, $\mathbf{u}(R) = \frac{1}{R^2}\mathbf{G}(R|R-\epsilon)$ is given by (18) with $I = J_\nu(K_L)$ (the equality is to within arbitrarily small ϵ/R). Obviously, this case may also be seen as a limit of vanishingly small optical penetration length α^{-1}.

4. Numerical treatment and the simulation results

4.1. Inverse Fourier transform

2D Fourier transform, discrete in circumferential number ν and continuous in frequency ω, has next been performed to obtain the sought acoustic response in the time-space domain $\{r, \theta, t\}$. The calculation has been made for the radial component of response at the surface, which is most usually the quantity measured by the laser ultrasonics technique. Denote its above-derived value in the transform domain as $u_r(R) = \widehat{u}_r(R, \nu, \omega)$, where $u_r(R)$ is given by Eq.(18). Fast Fourier Transform algorithm in frequency has been applied, with a small imaginary part added to ω ($\omega^* = \omega - \jmath\delta$ where $\delta > 0$, e.g. [12]) in order to avoid zeros of the dispersion equation $D = 0$, see (19) and the remark to (15). On taking Fourier transform in ν, the use has been made of the fact that, by (18), $\widehat{u}_r(r, \nu, \omega)$ is an even function of ν. As a result the sought response $u_r(R, \theta, t)$ is described by the formula

$$u_r(R, \theta, t) = \frac{e^{-\delta t}}{\pi} \int_{-\infty}^{\infty} \left\{ \sum_{\nu=0}^{\infty} \varepsilon_\nu \widehat{u}_r(R, \nu, \omega) cos(\nu\theta) \right\} e^{-\jmath\omega t} d\omega, \tag{20}$$

where ε_ν is the Neumann factor equal to 1 if $\nu = 0$ and to 2 otherwise.

The series in (20) is non-absolutely convergent [10]. Its convergence in ν depends on ω and hence so does the truncation order ν_m, which is supposed to satisfy the criterion

$$\frac{|\widehat{u}_r(R, \nu_m, \omega)|}{\sum_{\nu=0}^{\nu_m} |\widehat{u}_r(R, \nu, \omega)|} < \varepsilon, \tag{21}$$

where ε is the error bound. It is, however, known that usually the sum in question is mainly due to the first terms of the sum. For the frequency band used in our case, it was verified that taking a fixed truncation order ν_m such that keeps about 200 terms of the series ensures that the committed error does not exceed 10^{-5}.

4.2. Results of numerical simulations

Numerical simulations have been implemented for the opaque aluminium rod and the optically absorptive NG5 coloured glass rod, both of the diameter 5 mm. The values of their elastic and thermal constants are listed in Table 1 (the notations are explained below (1)). Assuming the laser wavelength 355 nm, the absorption length for NG5 coloured glass is $\alpha^{-1} = 0.8$ mm. The typical values of the laser beam-width $b = 0.15$ mm and of the pulse duration 5 ns are taken. The radial component of the surface displacement $u_r(R, \theta, t)$ is supposed to be detected at the point $(r, \theta) = (R, \pi)$ which is diametrically opposite to the generation point $(R, 0)$. Fourier transform in ω has been performed with the numerical parameter $\delta = 0.035$, see (20).

The results of numerical simulations are presented in Figs.1 and 2. The wave arrivals are marked by the dashed lines. They are identified and denoted according to the nomenclature proposed in [4] on the basis of a ray trajectory analysis. The notation nL and nT corresponds to the arrivals of longitudinal and transverse bulk modes, travelling along the straight diameter path and undergoing n normal reflections from the cylinder surface at the edge points. The notation nP and nS also corresponds to, respectively,

Table 1. Physical properties of Al and NG5 glass used in simulations.

	ρ [g cm^{-3}]	λ [GPa]	μ [GPa]	α_T [K^{-1}]	c_p [J kg^{-1} K^{-1}]
Al	2.69	56	26.5	2.5e-6	902
NG5	2.31	17.4	22.9	6.5e-6	700

the longitudinal and transverse bulk modes, but these ones arrive at the detection point after a broken path resulting from n oblique reflections without modal conversion. The wave arrivals, involving modal conversion at oblique reflections, are denoted by $mPnS_r$, where the subscript r is the number of round trips about the center made within the given path. The Rayleigh-wave arrival is denoted by R_r. The notation H_r is used for the wave arrival due to the head wave, i.e., due to the transverse mode radiated by the skimming longitudinal wave in the direction of the critical angle $\varphi_c = \arcsin\sqrt{2\mu/\left(\lambda+2\mu\right)}$.

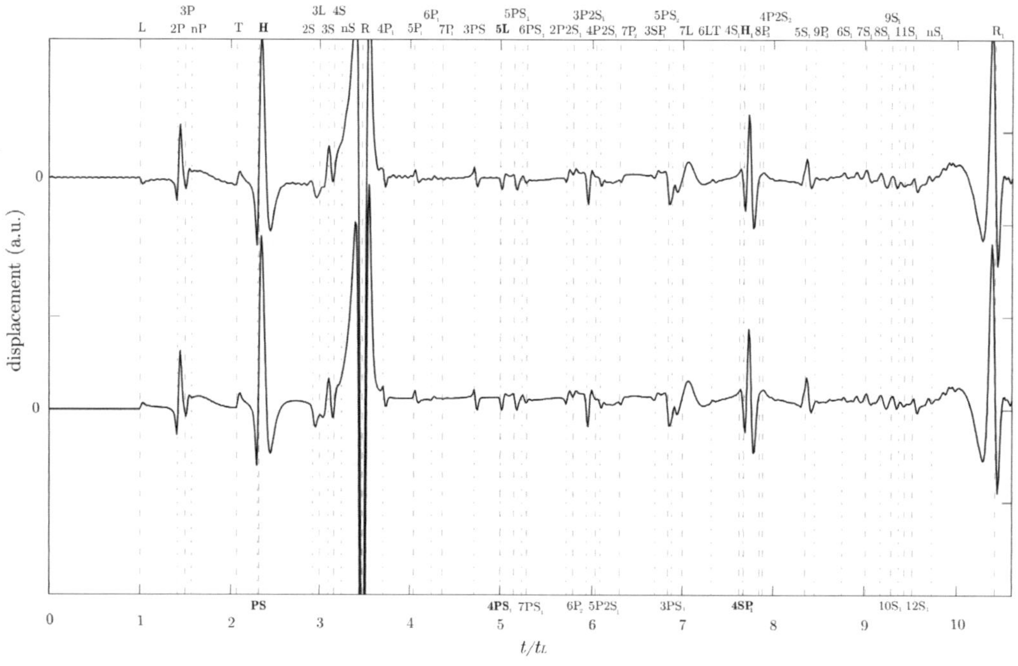

Figure 1. Acoustic response of the aluminium rod computed for the model of a buried source tending to the surface (upper curve) and for Scruby's model of a surface source (lower curve). The notations for wave arrivals are explained in text.

Fig.1 compares two simulations of the laser-generated acoustic response $u_r(R, \pi, t)$ for the opaque aluminium cylinder. The upper curve is computed by means of the developed procedure reduced to the model case such that assumes a radially concentrated buried source tending to the cylinder surface or, equally, a vanishingly small optical penetration length (see the closing remark in § 3.3). The lower curve is computed by means of Scruby's model of the surface dipole source [5]. Precise agreement of the arrival times and waveform shapes of both signals is observed.

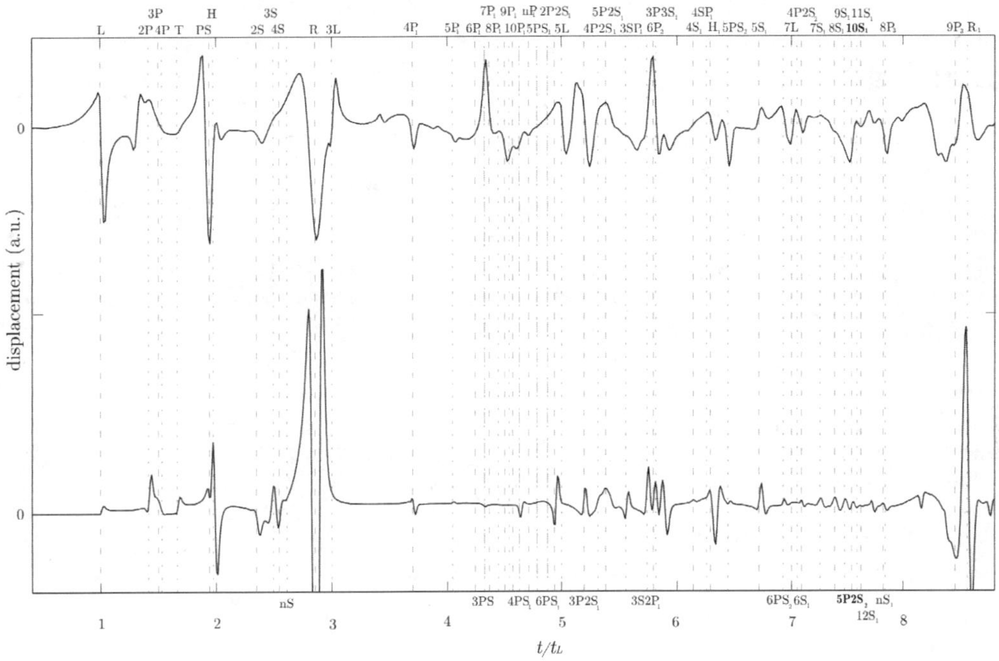

Figure 2. Acoustic response of the optically absorptive NG5 coloured glass rod. The upper curve is calculated for the radially distributed source taking into account optical penetration. The lower curve is calculated from Scruby's model of a surface source for opaque materials.

Fig.2 presents the numerical results for the optically absorptive NG5 coloured glass cylinder. The upper curve shows the acoustic response $u_r(R, \pi, t)$ computed from Eqs. (18)-(20) for the radially distributed source, which takes into account marked enough optical penetration of the given material ($\alpha R \approx 3$). This curve is again compared with the 'would-be' signal calculated for this case on the basis of Scruby's surface-source model (lower curve). It is seen that while the arrival times for both signals are obviously in good accordance, there is an evident discrepancy in the shape of the waveforms. For instance, the first longitudinal-wave arrival L on the upper curve has a characteristic bipolar broad shape, which is not observed on the lower curve. Another striking dissimilarity is that the bulk-wave and surface-wave arrivals on the upper curve have more or less similar amplitudes, which is in contrast to the predominant peaks of the first Rayleigh-wave arrivals R and R_1 on the lower curve. Altogether, the difference between these two curves highlights the effect of optical penetration, which Scruby's surface-source model is not intended to describe.

5. Conclusions

Acoustic response of an isotropic cylinder to a laser line pulse has been simulated in the framework of 2D semi-analytical model, which takes into account optical penetration into the bulk of optically absorptive cylinder material. The waveforms calculated for the aluminium rod with a vanishingly small penetration coincide, as expected, with those obtained from Scruby's model of a surface dipole source. At the same time, calculation for the glass rod with large enough penetration length has demonstrated an essential difference of the waveform shapes relatively to the signal calculated for the same material by way of the surface-source model.

The semi-analytical procedure developed in this paper paves the way to quantitative analysis of experimental results of laser ultrasonics generation in cylindrical samples made of optically absorptive materials. First experimental data have recently been obtained and will be reported elsewhere.

References

[1] Clorennec D, Royer D and Walaszek H 2002 *Ultrasonics* **40** 783 – 789
[2] Clorennec D and Royer D 2003 *Appl. Phys. Lett.* **82** 4608–4610
[3] Pan Y, Rossignol C and Audoin B 2003 *Appl. Phys. Lett.* **82** 4379–4381
[4] Pan Y, Rossignol C and Audoin B 2004 *J. Acoust. Soc. Am.* **115** 1537–1545
[5] Scruby C B and Drain L E 1990 *Laser Ultrasonics Technique & Applications* (Adam Hilger)
[6] Royer D 2001 *Ultrasonics* **39** 345 – 354
[7] Arias I and Achenbach J D 2003 *Int. J. Solids Struct.* **40** 6917 – 6935 special issue in Honor of George J. Dvorak
[8] Dubois M, Enguehard F and Bertrand L 1994 *Phys. Rev. E* **50** 1548–1551
[9] Hecht E 2002 *Optics* (San Fransisco: Addison Wesley)
[10] Rousselot J L 1987 *La Diffusion Acoustique par des Cibles Elastiques de Formes Géométrique Simple: Théories et Expériences* ed Gespa N (Paris: CEDOCAR) chap 21
[11] Kamke E 1971 *Differentialgleichungen: Lösungsmethoden und Lösungen* (New-York: Chelsea)
[12] Weaver R L, Sachse W and Kim K Y 1996 *J. Appl. Mech.* **63** 337–346

Anglo–French Physical Acoustics Conference 2009 IOP Publishing
Journal of Physics: Conference Series **195** (2009) 012008 doi:10.1088/1742-6596/195/1/012008

The effects of nonlinear wave propagation on the stability of inertial cavitation

D Sinden, E Stride & N Saffari

Mechanical Engineering, Roberts Building, Torrington Place, London, WC1E 7JE, United Kingdom

E-mail: d.sinden@ucl.ac.uk

Abstract. In the context of forecasting temperature and pressure fields generated by high-intensity focussed ultrasound, the accuracy of predictive models is critical for the safety and efficacy of treatment. In such fields 'inertial' cavitation is often observed. Classically, estimations of cavitation thresholds have been based on the assumption that the incident wave at the surface of a bubble is the same as in the far-field, neglecting the effect of nonlinear wave propagation. By modelling the incident wave as a solution to Burgers' equation using weak shock theory, the effects of nonlinear wave propagation on inertial cavitation are investigated using both numerical and analytical techniques. From radius-time curves for a single bubble, it is observed that there is a reduction in the maximum size of a bubble undergoing inertial cavitation and that the inertial collapse occurs earlier in contrast with the classical case. Corresponding stability thresholds for a bubble whose initial radius is slightly below the critical Blake radius are calculated, providing a lower bound for the onset of instability. Bifurcation diagrams and frequency-response curves are presented associated with the loss of stability. The consequences and physical implications of the results are discussed with respect to the classical results.

1. Introduction

For materials with a nonlinear stress-strain relationship, such as tissue [1], the point of maximum compression for a wave may propagate faster than the point of maximum rarefaction, leading to a distortion in the wave profile and the redistribution of energy from the fundamental harmonic frequency to higher harmonics. The pressures associated with therapeutic high intensity focused ultrasound (HIFU) may be high enough for the effects of nonlinear wave propagation to be significant [2]. As high frequency components are absorbed more easily than those with lower frequencies, nonlinear wave propagation contributes to increased absorption, enhanced heating and a subsequent shift in the focal point of targeted ultrasonic beam, potentially damaging healthy tissue. As well as providing greater predictive accuracy, knowledge of nonlinear wave propagation will enable increased accuracy in calibration using the received signal generated by bubble oscillations [3] and thus greater accuracy in treatment procedures. However, in this paper the implications of nonlinear wave propagation on inertial cavitation are investigated and stability criteria re-derived in an attempt to reconcile theory and experiment.

The Rayleigh-Plesset equation [4] is a nonlinear equation which determines the size of a spherical bubble subject to a varying pressure field. Wave propagation from the far-field to the surface of the bubble is generally assumed to be linear, yet this does not correspond with experimental observations in the context of high intensity focused ultrasound. Moss [5] attempted to incorporate the effects of boundary conditions and global compressibility/local incompressibility into the Rayleigh-Plesset equation but only considered linear wave propagation. The equation derived was identical to the classical Rayleigh-Plesset equation if the far-field pressure is replaced by an attenuated pressure measured within 25 bubble radii of the bubble surface. In this paper the effect of distortion rather than attenuation of the wave profile and its effect on the stability of oscillations will be investigated.

Lauterborn and co-workers [4,6] and Smerka [7] showed experimentally and numerically that bubble oscillations undergo a sequence of period-doubling bifurcations leading to unstable quasi-periodic (chaotic) oscillations. The period doubling route to chaos occurs through a succession of saddle-node bifurcations of subharmonic periodic orbits. Homoclinic bifurcations are the limit of a countable sequence of subharmonic saddle-node bifurcations and thus provide an insight into the parameters regimes at which unbounded growth occurs.

Mel'nikov's method provides a measure of the distance between the stable and unstable manifolds of a periodically perturbed system [8]. If the manifolds intersect transversely once, they will intersect transversely infinitely many more times. The transverse intersections can be represented by Smale horseshoes, through which the Smale-Birkhoff theorem give an elegant description of sensitivity to initial conditions and the resulting chaotic oscillations. Hence the Mel'nikov criteria marks the first event in a sequence of events leading to unbounded bubble growth and as such provides a valuable lower bound threshold.

The first application of Mel'nikov's method to cavitation was performed by Chang and Chen [9] on the effect of viscosity on the Hamiltonian structure. Szeri [10] computes the first-order splitting for the Rayleigh-Plesset equation numerically. Harkin [11] performs Mel'nikov analysis analyically on bubbles whose initial radius is slightly smaller than Blake critical radius. Using matched perturbation analysis, Harkin derives a second order normal form for the Rayleigh-Plesset equation. The normal form is a damped driven oscillator. An escape velocity, like the static Blake criterion, provides an upper bound for when unbounded growth will occur, whereas Mel'nikov's method provides a lower bound for when the transition to chaos and unbounded growth may occur. The fate of bubbles whose initial conditions lie in the intermediate region between the Mel'nikov and Blake thresholds can be computed by a transport-type processes [12]. A Bernoulli shift map on two symbols has already been constructed numerically from a bifurcation diagram for the full Rayleigh-Plesset equation [13] without explicit inference to sensitivity to initial conditions. In each application of Mel'nikov's method the incident pressure wave was sinusoidal [7,9,11,10]. In this paper the analysis is performed for nonlinear waves.

The outline of this paper is as follows, in section 2 nonlinear wave propagation is considered and wave profiles derived. Then in section 3 the Rayleigh-Plesset equation is introduced and the effects of nonlinear wave propagation investigated. In section 4 Mel'nikov analysis is performed for nonlinear wave propagation, providing an improved measure of the values at which quasi-periodic oscillation and unbounded growth may occur. Finally, conclusions and implications are discussed in section 5.

2. Nonlinear Wave Propagation

There is no universally accepted system of partial differential equations for the modelling of ultrasound propagation in biological tissue. Perhaps the best known is the Khokhlov-Zabolotskaya-Kuznetsov (KZK) equation. The KZK equation is a parabolic wave equation which includes the effects of diffraction, absorption and nonlinearity of the directed beams [3]. The KZK equation for an axi-symmetric beam which propagates in the r direction is written in terms of the acoustic pressure $p(r,t)$ as

$$\frac{\partial^2 p}{\partial r \partial t'} = \frac{c_0}{2}\nabla_{\perp}^2 p + \frac{D}{2c_0^2}\frac{\partial^3 p}{\partial t'^3} + \frac{\beta}{2\rho c_0^3}\frac{\partial^2 p^2}{\partial t'^2} \tag{1}$$

where c_0 is the initial wave speed, ρ is the density of the medium, β is the standard nonlinearity coefficient given by $\beta = 1 + B/2A$, where B/A is the standard nonlinearity parameter, D is the sound diffusivity and $t' = t - r/c_0$ is the retarded time variable. The Laplacian is taken with respect to the transverse coordinates. The sound diffusivity is given by

$$D = \frac{1}{\rho}\left(\mu_b + 4\mu_s/3 + \kappa\left(1/c_v - 1/c_p\right)\right)$$

where μ_b and μ_s are the bulk and shear viscosities respectively, $1 \leq \kappa \leq \gamma$ the polytropic exponent, and c_v and c_p are the specific heat capacities at constant volume and pressure respectively. If $\kappa = 1$ the system is isothermal, if $\kappa = \gamma \equiv c_p/c_v$ the system is adiabatic. In the study of cavitation it will be assumed that the bubble is essentially stationary over the time scales considered and that the distance from the surface of the bubble to the shock front is constant, so that r is a fixed value at the surface of the bubble. The equation (1) can be integrated with respect to retarded time to give

$$\frac{\partial p}{\partial r} = \frac{c_0}{2}\int_{-\infty}^{t'} \nabla_{\perp}^2 p\left(s,t\right)\,\mathrm{d}s + \frac{D}{2c_0^2}\frac{\partial^2 p}{\partial t'^2} + \frac{\beta}{2\rho c_0^3}\frac{\partial p^2}{\partial t'}. \tag{2}$$

Discarding the effects of diffraction Burgers', or alternatively the Burgers-Hopf, equation is recovered

$$\frac{\partial p}{\partial r} = \frac{D}{2c_0^2}\frac{\partial^2 p}{\partial t'^2} + \frac{\beta}{2\rho c_0^3}\frac{\partial p^2}{\partial t'}. \tag{3}$$

If $D = 0$ the fluid is inviscid, and the so-called lossless Burgers equation is recovered.

From a given driving pressure, $f\left(t\right)$, the incident pressure wave, $p\left(t\right)$, can be expressed using weak shock theory. The location of a shock is determined by the Rankine-Hugoniot relation defining the conservation of flux. The areas enclosed by the multi-valued solution to the left and right of the shock are equal. Thus, by this symmetry, the shock is positioned at the zero of the linear solution. For a sinusoidal driving pressure of magnitude P and frequency ω, i.e.

$$f = P\sin\left(\omega t\right), \tag{4}$$

a Fourier expansion of the solution to the lossless Burgers equation yields the Bessel-Fubini or alternately the Fubini solution

$$p = \frac{2P}{r_s}\sum_{n=1}^{\infty}\frac{1}{n}J_n\left(nr_s\right)\sin\left(\omega_n t'\right), \tag{5}$$

where $\omega_n = n\omega$, J_n is an nth-order Bessel functions of the first kind and r_s is the normalised distance to a shock given by

$$r_s = \frac{r}{r_c} \quad \text{where} \quad r_c = \frac{\rho c_0^3}{\beta\omega P}$$

is the location of the shock.

Beyond the shock, weak shock theory can once again be employed to find a solution, however, the resulting analytical solution takes an integral form. An asymptotic solution, valid for, say $r_s > 3$, is

$$p = \frac{2P}{1+r_s}\sum_{n=1}^{\infty}\frac{1}{n}\sin\left(\omega_n t'\right). \tag{6}$$

which can be expressed in the time domain as

$$p = \begin{cases} -\dfrac{P\left(\omega t' + \pi\right)}{1 + r_s} & \text{for} \quad -\pi < \omega t' < 0, \\ -\dfrac{P\left(\omega t' - \pi\right)}{1 + r_s} & \text{for} \quad 0 < \omega t' < \pi. \end{cases}$$

A general solution to the Burgers'-Hopf equation for a sinusoidal driving pressure (4), derived by transforming the nonlinear equation into a linear diffusion equation via the Hopf-Cole transformation, is given by

$$p = -\frac{4P}{\kappa} \frac{\sum_{n=1}^{\infty} n \left(-1\right)^n I_n \left(\Gamma/2\right) e^{-n^2 r_s/\Gamma} \sin\left(\omega_n t'\right)}{I_0\left(\Gamma/2\right) + 2\sum_{n=1}^{\infty} \left(-1\right)^n I_n\left(\Gamma/2\right) e^{-n^2 r_s/\Gamma} \cos\left(\omega_n t'\right)} \tag{7}$$

where I_n is an nth-order modified Bessel functions of the first kind and Γ is the Gol'dberg number which relates the importance of nonlinear effects against dissipative effects as

$$\Gamma = \frac{2\beta P}{D\rho\omega}.$$

If $\Gamma \gg 1$ then the relative effects of nonlinearity are strong, if $\Gamma < 1$ the relative effects of nonlinearity are weak.

Unlike the Fubini solution (5), for strong nonlinearity, i.e. $\Gamma \gg 1$, the solution converges slowly but far from the shock, i.e. again say $r_s > 3$, asymptotic analysis yields the more usable formulation

$$p = \frac{2P}{\Gamma} \sum_{n=1}^{\infty} \frac{\sin\left(\omega_n t'\right)}{\sinh\left(n\left(1 + r_s\right)/\Gamma\right)} \tag{8}$$

called the Fay solution. The Fay solution can be expressed as the periodic function

$$p = \frac{P}{1 + r_s}\left(-\omega t' + \pi \tanh\left(\frac{\pi \Gamma \omega t'}{2\left(1 + r_s\right)}\right)\right) \qquad -\pi \le \omega t' \le \pi.$$

In the lossless limit as viscosity vanishes, that is as $\Gamma \to \infty$, the Fay solution recovers the Fourier expression for a sawtooth function (6). The Fubini (5) and Fay (8) solutions may at first seem incompatible but each holds in a different region of the flow; the Fubini solution close to the source as a shock begins to form and the Fay solution far from the source as a shock begins to decay. Blackstock [14] shows that the true solution to the lossless Burgers equation over the entire domain is simply the sum of the two solutions (5) and (6). In the near-field the Fubini solution is dominant, in the far-field the sawtooth solution is dominant. In [15] exact solutions of Burgers' equation from the Cole-Hopf transformation were computed numerically and contrasted against the analytical solutions and showed good agreement between the sets of solutions unless in the immediate neighbourhood of the shock. The formation of a shock will occur in a region of peak negative pressure and hence is a prime site for the nucleation of cavities. Thus the forthcoming analysis will be performed either before a shock or significantly after, so that only the effect of nonlinear wave propagation on pre-existing cavities is studied.

Figure 1 presents nonlinear wave profiles (5) and a linear wave profile (4) illustrating the distortion due to nonlinear propagation of waves with equal amplitude. Throughout this paper comparisons between linear and nonlinear waves of equal amplitude are studied.

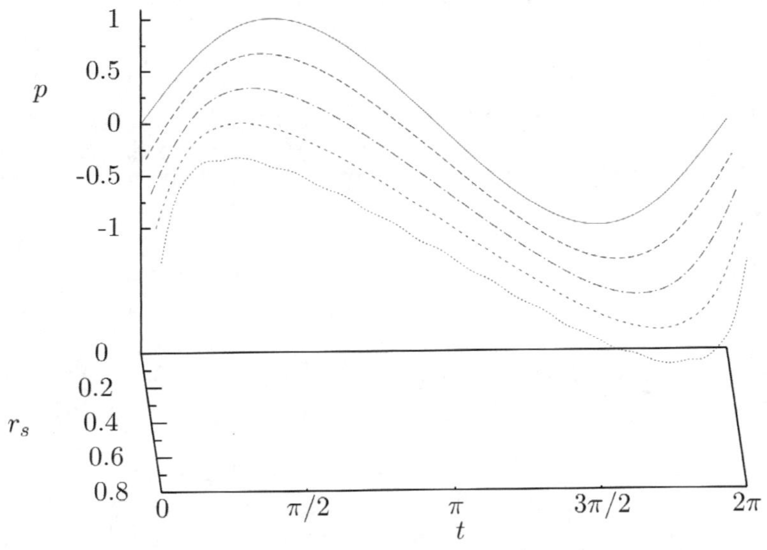

Figure 1. Profiles of incident waves for linear and nonlinear wave propagation as the distance from a shock front decreases. The nonlinear waves are modeled by the first twenty terms of the (normalised) Fubini solution. The linear wave is modelled when $r = 0$ (red,solid), nonlinear waves are $r_s = 1/5$ (blue, dashed), $r_s = 2/5$ (blue, dot-dashed), $r_s = 3/5$ (blue, short-dashed) and $r_s = 4/5$ (blue, dot).

3. Rayleigh-Plesset Equation

The Rayleigh-Plesset equation is an ordinary differential equation which models the oscillations of a spherical bubble of radius $R = R(t)$ whose centre is stationary. The Rayleigh-Plesset equation can be derived by balancing the energy supplied to the bubble by the incident pressure and the surrounding fluid and the kinetic energy of the bubble oscillations [16] yielding

$$\rho\left(R\ddot{R} + \frac{3}{2}\dot{R}^2\right) = p_g(R) + p_v - p_\infty + p(t) + \frac{2\sigma}{R} + \frac{4\mu\dot{R}}{R} \tag{9}$$

where p_g is the internal pressure of the gas in the bubble given by the hard-core van der Waals relationship

$$p_g(R(t)) = \left(p_\infty - p_v + \frac{2\sigma}{R_0}\right)\left(\frac{R_0^3 - a^3}{R(t)^3 - a^3}\right)^\kappa \tag{10}$$

with R_0 the equilibrium radius, a the van der Waals hard-core radius, p_v the vapour pressure, p_∞ is the ambient pressure, σ is the surface tension and μ viscosity. The gas is assumed to be ideal as the internal pressure is a function of the bubble radius only. The forcing pressure will

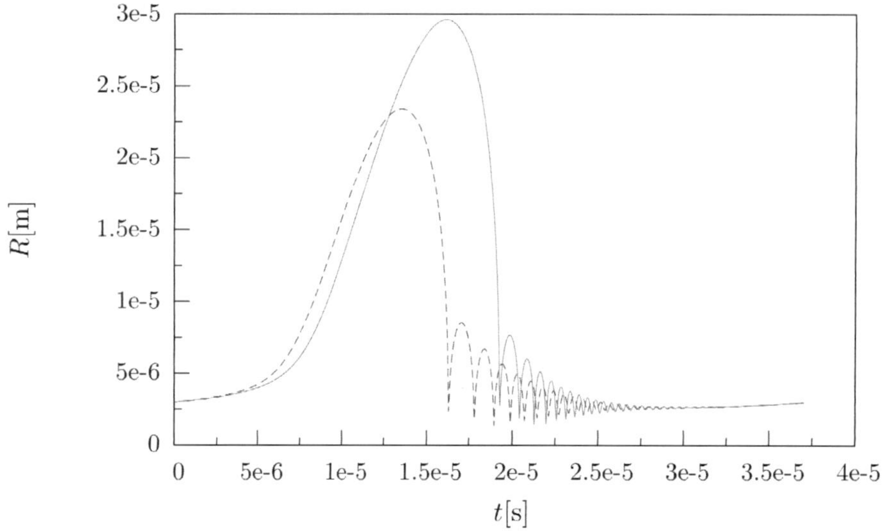

Figure 2. Contrasting profiles of inertial cavitation for linear (red, solid) and nonlinear (blue, dashed) wave propagation modeled by the first ten terms of the Fubini solution. $R_0 = 0.6\mu\text{m}$, subject to a wave with frequency $\omega = 2\pi \cdot 26\text{kHz}$, amplitude of driving pressure $P = 1.36\text{atm}$, normalised distance to the shock $r_s = 1/20$ ambient pressure $p_\infty = 1\text{atm}$, surface tension $\sigma = 0.073\text{kgm}^{-2}$, viscosity $\mu = 10^{-3}\text{kgm}^{-1}\text{s}^{-1}$, density $\rho = 1000\text{kgm}^{-3}$, polytropic exponent $\kappa = 5/3$ and hard-core radius $a = R_0/8.85$.

take the form

$$p(t) = \sum_{n=1}^{\infty} P_n \sin(\omega_n (t + t_0))$$

where P_n are the Fourier terms of a solution to a Burgers' equation, such as those given by the Fubini (5) or the Fay (8) solutions.

Figure 2 contrasts the effects of nonlinear and linear wave propagation. It is clear that for nonlinear wave propagation inertial cavitation occurs before and at a smaller maximum radius than for linearly propagated waves. This has two significant effects, firstly as the collapse occurs at a small maximum radius there is a diminished chance of shape instability [17] and secondly, as the collapse occurs earlier more after-bounces can occur so that the bubble returns to a stable initial radius before the next cycle begins. Note that the period of the after-bounces is the same for the nonlinear and linear wave since the after-bounces occur at the natural frequency of the Rayleigh-Plesset equation which is independent of the applied pressure.

4. Mel'nikov Analysis

Considering an unforced bubble and let $R(t) = R_0(1 + x)$. When $x \ll 1$ then to first order the nondimensional Rayleigh-Plesset equation yields the simple harmonic equation

$$\ddot{x} + \omega_0 x = 0 \quad \text{where} \quad \omega_0^2 = \frac{1}{\rho}\left(\frac{3\kappa(p_\infty - p_v)R_0}{R_0^3 - a^3} + 2\sigma\left(\frac{3\kappa}{R_0^3 - a^3} - \frac{1}{R_0^3}\right)\right). \tag{11}$$

The natural frequency ω_0 is used to nondimensionalise time by $\tau = \omega_0 t$ so that the new time-like variable will be a function of the perturbation parameter. Note that when $a = 0$ and $\kappa = 1$ then the frequency given by Harkin [11, Eq. (12)] is recovered. In this section Harkin's analysis is followed as it gives an analytical expression for the transverse intersections which allows for direct comparison between linear and nonlinear wave propagation.

The static Blake threshold pressure is the point at which the internal pressure $p_v + p_g$ is equal to the external pressure $p_l + 2\sigma/R$, thus for internal pressures larger than this threshold unbounded growth will occur. The equilibrium pressure exerted on the bubble surface by the liquid, p_l, is given by

$$p_l = p_g + p_v - \frac{2\sigma}{R}. \tag{12}$$

Now perturb the equilibrium radius by $R(t) = R_0(1 + \epsilon x(\tau))$ with ϵ a small parameter given by

$$\epsilon = 2(1 - R_0/R_c)$$

where $R_c = R_c\left(\tilde{G}, \sigma, \kappa, a\right)$ is the Blake critical radius, found as the stationary solutions to the unforced Rayleigh-Plesset equation

$$\frac{3\kappa\tilde{G}R^4}{\left(R^3 - a^3\right)^{(\kappa+1)}} = 2\sigma \quad \text{where} \quad \tilde{G} = \left(p_\infty - p_v + \frac{2\sigma}{R_0}\right)\left(R_0^3 - a^3\right)^\kappa. \tag{13}$$

Unfortunately if $a \neq 0$ then there is no simple analytical expression for the critical Blake radius. In the isothermal case the critical radius can be found explicitly as the solution to the cubic equation

$$R^3 - \sqrt{\frac{3\tilde{G}}{2\sigma}}R^2 - a^3 = 0.$$

It is straightforward to show from the discriminant of the cubic equation when $a \geq 0$ and $\tilde{G} \geq 0$ that the equation will have one real solution and a pair of ignorable complex conjugate solutions. In the case of no hard-core radius, i.e. $a = 0$, then

$$R_c = \sqrt[3\kappa-1]{\frac{3\kappa\tilde{G}}{2\sigma}} \tag{14}$$

so that the critical value for the liquid pressure (12) is then

$$p_c = p_v - \sqrt[3\kappa-1]{\frac{(2\sigma)^{3\kappa}}{3\kappa\tilde{G}}}\left(1 - \frac{1}{3\kappa}\right). \tag{15}$$

On combining (14) and (15) the Blake critical radius is then given in the familiar form by

$$R_c = \frac{2\sigma}{(p_v - p_c)}\left(1 - \frac{1}{3\kappa}\right). \tag{16}$$

Note that

$$\begin{aligned} p_\infty - p_v &= \frac{2\sigma}{3R_0}\left(\frac{R_c}{R_0}\right)^2 - \frac{2\sigma}{R_0} \\ &= \frac{2\sigma}{3\kappa R_0}\left(1 - 3\kappa + \epsilon + \frac{1}{2}\frac{3\kappa}{3\kappa - 1}\epsilon^2 + \mathcal{O}\left(\epsilon^3\right)\right), \end{aligned} \tag{17}$$

and similarly

$$p_c - p_v = \frac{2\sigma(3\kappa - 1)}{3\kappa R_0}\left(1 - \frac{\epsilon}{2}\right)^{-1} \tag{18}$$

so the critical pressure p_c and the ambient pressure p_∞ in general differ by terms $\mathcal{O}(\epsilon)$ but when isothermal $\mathcal{O}(\epsilon^2)$. Thus when the equilibrium radius is close to the critical radius and the ambient pressure is close to the critical pressure, the natural frequency can be expressed as

$$\omega_0^2 = \frac{2\sigma\epsilon(3\kappa - 1)}{\rho R_0^3}. \tag{19}$$

When the perturbation is applied to the Rayleigh-Plesset equation (9) in this regime both the $\mathcal{O}(1)$ and $\mathcal{O}(\epsilon)$ terms are zero *if and only if the system is isothermal*, i.e. $\kappa = 1$. Thus, when isothermal to $\mathcal{O}(\epsilon^2)$ the governing equation is a Helmholtz oscillator

$$\ddot{x} + 2\zeta\dot{x} + x(1-x) = \sum_{n=1}^{N} A_n \sin(\Omega_n(\tau + \tau_0)), \tag{20}$$

where the overdot represents the derivative with respect to the nondimensional time τ and

$$\zeta = \mu\sqrt{\frac{2}{\epsilon\sigma\rho R_0}}, \quad A_n = \frac{P_n R_0}{2\sigma\epsilon^2} \quad \text{and} \quad \Omega_n = \omega_n\sqrt{\frac{\rho R_0^3}{2\sigma\epsilon}}$$

are constant terms of $\mathcal{O}(1)$ when ϵ is small for bubbles whose initial radius is of order microns driven by frequencies in the megahertz range [11]. Here ζ is the nondimensional damping, Ω_n are harmonics of the nondimensional frequency, A_n the nondimensional Fourier components of the applied pressure and τ_0 is the phase of the incident wave. Note that the series expansion is truncated to N terms in order to disregard terms which are greater than $\mathcal{O}(\epsilon^2)$.

When forcing and viscosity are rescaled as small parameters, $\zeta \mapsto \varepsilon\zeta$ and $f \mapsto \varepsilon f$, which both destroy the integrable Hamiltonian structure then the Mel'nikov integral can be calculated. Let the ε-perturbed system be given by $\dot{\boldsymbol{x}} = \boldsymbol{f}_0(\boldsymbol{x}) + \varepsilon\boldsymbol{f}_1(\boldsymbol{x},\tau)$ with $\boldsymbol{x} = (x,y)^T = (x,\dot{x})^T$ and $\boldsymbol{f}_1$ an Ω-periodic function. Explicitly the vector field is given by

$$\begin{aligned}\dot{x} &= y, \\ \dot{y} &= x(x-1) + \varepsilon\left(\sum_{n=1}^{N} A_n \sin(\Omega_n\tau) - 2\zeta y\right).\end{aligned}$$

The unperturbed system, $\varepsilon = 0$, admits a homoclinic orbit $\boldsymbol{\gamma}$ emanating from the saddle at $(1,0)$ of the form

$$\boldsymbol{\gamma}(\tau) = \frac{1}{2}\left(\tanh^2\left(\frac{\tau}{2}\right) - 1, 3\tanh\left(\frac{\tau}{2}\right)\operatorname{sech}^2\left(\frac{\tau}{2}\right)\right).$$

The first order Mel'nikov integral at the homoclinic energy level h can simply be calculated using Cauchy's residue theory as

$$\begin{aligned}\mathcal{M}(\tau_0) &= \int_{-\infty}^{+\infty} \boldsymbol{f}_0(\boldsymbol{x}) \wedge \boldsymbol{f}_1(\boldsymbol{x}, \tau + \tau_0)\, \mathrm{d}\tau \\ &= 6\pi\sum_{n=1}^{\infty} \frac{A_n\Omega_n^2}{\sinh(\Omega_n\pi)}\cos(\Omega_n\tau_0) - \frac{12\zeta}{5}.\end{aligned} \tag{21}$$

Due to the summation, it is not possible to formulate an explicit condition for simple zeros, but instead perform a calculation numerically. For both the Fubini and the Fay solutions it is simple to show numerically that the Mel'nikov integral has simple zeroes for larger A_n than in the case of linear wave propagation.

For the Fubini solution as $r_s \to 0$ the linear result is recovered and no summation is required, that is as $r_s \to 0$ so $P_1 \to P$, $P_j \to 0$ for all $j = 2, 3, \ldots$. As the distance towards the shock decreases, i.e. r_s increases from zero, so a higher P or larger R_0 is necessary in order for the Mel'nikov integral to have simple zeroes.

For the Fay solution higher pressures are required further from the shock. Also, higher pressures are required for materials with low Gol'dberg numbers than for materials with high Gol'dberg numbers.

A Poincaré section can be constructed which is topologically conjugate to a Bernoulli shift map on two symbols so that the likelihood of, say, $R_{\max}$ being greater or smaller than the previous cycle is as random as the toss of a coin. Thus, for sufficiently small ε, chaotic bubble oscillations will occur in the vicinity of γ. From the simplified normal form, the effect of nonlinear wave propagation implies that a cascade to chaos and unbounded bubble growth will occur at higher driving pressures P or larger initial radii R_0 than for linear wave propagation. As $\Omega \to \infty$ then $\mathcal{M} \to 12\zeta/5$ and higher order contributions vanish, so that the stable and unstable manifolds stay disjoint. However, as $\Omega \to 0$ then $\mathcal{M} \to 12\zeta/5$, but now second order terms will affect the threshold criteria [18]. Indeed, numerical analysis suggests that higher order contributions must be taken into account in this regime. Furthermore, the authors [18] state that higher-order Mel'nikov analysis, i.e. a second aproximation of the separation distance between the stable and unstable manifolds is always necessary when considering nonlinear wave propagation due to the interactions of the differing harmonic excitations. Note that the classical first order Mel'nikov integral will typically overestimate the threshold by not including higher order contributions.

It is simple to show that in the absence of (nondimensional) viscous damping, i.e. $\zeta = 0$, that generically the Mel'nikov integral will have simple zeros for all (non-zero) parameter values. The effect of viscosity actually reduces the likelihood of violent cavitation, a result also found by Szeri [10]. No such analysis has yet been performed in the visco-elastic case but it is currently under investigation.

The bifurcation diagrams presented in figures 3 and 4 were computed for the Rayleigh-Plesset equation (9) with an adaptive, explicit Runge-Kutta method of order $(8, 5)$ with error control due to Dormand and Prince [19]. In order to ignore transient behaviour the first fifty cycles were disregarded and the next fifty cycles plotted. Figure 3 generalises the inferences from the radius-time profiles displayed in figure 2. Subfigure 3(a) illustrates that the maximum amplitude of the bubbles under forcing from nonlinear wave propagation is smaller than the maximum amplitude of bubbles under forcing from linear wave propagation. By defining ξ as the phase of the minimum radius $R_{\min} = R(t_{\min})$ per acoustic cycle [13]

$$\xi = (t_{\min} - t) / \Omega$$

subfigure 3(b) illustrates that inertial cavitation occurs earlier in each acoustic cycle for nonlinear waves than for linear waves. Both bifurcation diagrams in figure 3 show the same regions of developing instability for linear waves (illustrated by the dotted lines) as both solution measures determine Poincaré sections tangential to the motion of the cavity, i.e. when $\dot{R} = 0$.

Figure 4 shows a bifurcation diagram beyond the critical threshold and figure 5 then shows a selection of associated radius-time profiles illustrating the cascade to chaos through a period doubling bifurcation. Subfigure 5(a) shows a stable one-period radius-time profile at $P = 1.30$atm, which undergoes a period doubling bifurcation so that at $P = 1.35$atm in subfigure 5(b) the profile is two-periodic. For $P = 1.375$atm as subfigure 5(c) illustrates the radius-time profile is four-periodic. By $P = 1.40$atm the pressure has passed an accumulation point of period doubling bifurcations and as subfigure 5(d) illustrates the bubble oscillations are chaotic. Note that the pressure intervals between periodic cycles decreases as the system approaches chaotic oscillations, characteristic of the cascade to chaos. From a clinical perspective, when the duty cycle maybe in the order of seconds rather than micro seconds, accurate predictions for the heat deposition on tumour sites over all but the shortest time scales can not be inferred from cavitation activity when the driving pressure is beyond the threshold.

Bifurcation diagrams of linear, see for example [13, figure 5], and nonlinear wave propagation, such as in figure 4, at equal driving pressures far beyond the threshold are significantly different, with differing regions of stability, indicating that beyond the threshold predictions based on inaccurate wave propagation models will be inaccurate for all but the shortest time.

5. Conclusion

In this paper the effect of nonlinear wave propagation of high intensity focused ultrasound on the stability of cavitating bubbles has been investigated numerically and analytically. The effect of nonlinear wave propagation redistributes energy from the primary harmonic to higher harmonics. From this two significant conclusions can be drawn; firstly the maximum bubble radius is reduced. This is clearly illustrated by the bifurcation diagram in subfigure 3(a). Thus the likelihood of shape instability is reduced. Secondly, inertial cavitation occurs earlier in each cycle, as illustrated by the bifurcation diagram in subfigure 3(b). The earlier onset of collapse allows for the bubble to return to an equilibrium radius and for inertial cavitation to reoccur.

Amongst the many threshold criteria applied to cavitation, it is worthwhile emphasising that although the Mel'nikov criteria does not exactly correspond to the escape boundary whereby bubbles increase without bound, it initiates the penetration of escaping tongues into the safe basin and therefore constitutes the first event in a well known sequence of complicated events which leads to unbounded bubble growth [20]. Thus the Mel'nikov criteria provides a lower bound threshold value. In the context of therapeutic applications, where safety is paramount, such a criterion is of great value.

In therapeutic applications a cloud of bubbles, comprised of many thousands of bubbles of differing equilibrium radii and resonance frequencies will be present, rather than a single, isolated bubble. Thus it could be assumed that a stability criterion for a single bubble is of limited use. However, the stability of a single bubble does provide insight into the stability of entire bubble clouds: experimental evidence [21] suggests that entire bubble clouds undergo period doubling cascade to chaos, not just the subset of bubbles which satisfy the Mel'nikov criterion. It is believed that the interaction between the bubbles results in an averaged behaviour. Indeed, numerical and experimental of the dimension of the attractor in phase space are remarkably low, between 2 and 2.5, indicating that the number of relevant degrees of freedom in the system is also low [22, 23].

In many applications of therapeutic ultrasound the tissue will have distributed inhomogeneities, leading to dissipation through diffraction, but it is conjectured that material nonlinearity is of greater significance than material inhomogeneity [24], although this subject is currently under investigation.

Acknowledgments

This work was supported under grant EP/F025750/1 and is gratefully acknowledged. The authors would like to thank S. Martynov, G. Vilensky and P. Gélat for insightful comments and discussions.

References

[1] Picinbono G, Delingette H and Ayache N 2001 Nonlinear and anisotropic elastic soft tissue models for medical simulation *Robotics and Automation, 2001. Proceedings 2001 ICRA. IEEE International Conference on* vol 2 (IEEE)
[2] Muir T G and Carstensen E L 1980 *Ultrasound Med. Biol.* **6** 345–57
[3] Humphrey V F 2000 *Ultrasonics* **38** 267–272
[4] Lauterborn W 1976 *J. Acoust. Soc. Am.* **59** 283–293
[5] Moss W C 1997 *J. Acoust. Soc. Am.* **101** 1187–1190

[6] Lauterborn W and Cramer E 1981 *Phys. Rev. Lett.* **47** 1445–1448
[7] Smereka P, Birnir B and Banerjee S 1987 *Phy. Fluids* **30** 3342–3350
[8] Guckenheimer J and Holmes P J 1983 *Nonlinear Oscillations, Dynamical systems, and Bifurcations of Vector Fields* (*Applied Mathematical Sciences* no 42) (New York, NY: Springer-Verlag)
[9] Chang H C and Chen L H 1986 *Phy. Fluids* **29** 3530–3536
[10] Szeri A J and Leal L G 1991 *Phy. Fluids* **3** 551–555
[11] Harkin A, Nadim A and Kaper T J 1999 *Phy. Fluids* **11** 274–287
[12] Kang I S and Leal L G 1990 *J. Fluid Mech.* **218** 41–69
[13] Simon G, Cvitanovic P, Levinsen M T, Csabai I and Horvath A 2002 *Nonlinearity* **15** 25–43
[14] Blackstock D T 1966 *J. Acoust. Soc. Am.* **39** 1019–1026
[15] Mitome H 1989 *J. Acoust. Soc. Am.* **86** 2334–2338
[16] Leighton T G 1994 *The Acoustic Bubble* (London: Academic Press)
[17] Brenner M P, Lohse D and Dupont T F 1995 *Phys. Rev. Lett.* **75** 954–957
[18] Lenci S and Rega G 2004 *Math. Probl. Eng.* **2004**(2) 145–168
[19] Hairer E, Nørsett S P and Wanner G 1993 *Solving ordinary differential equations I: Nonstiff problems* (*Springer Series in Computational Mathematics* no 8) (Berlin: Springer-Verlag)
[20] Thompson J M T 1989 *Proc. R. Soc. Lond. A* **421** 195–225
[21] Lauterborn W and Koch A 1987 *Phys. Rev. A* **35** 1974–1976
[22] Lauterborn W and Holzfuss J 1986 *Phys. Letts. A* **115** 369–372
[23] Holt R G, Gaitan D F, Atchley A A and Holzfuss J 1994 *Phys. Rev. Lett.* **72** 1376–1379
[24] Duck F A 1990 *Physical Properties of Tissue: A Comprehensive Reference Book* (London: Academic Press)

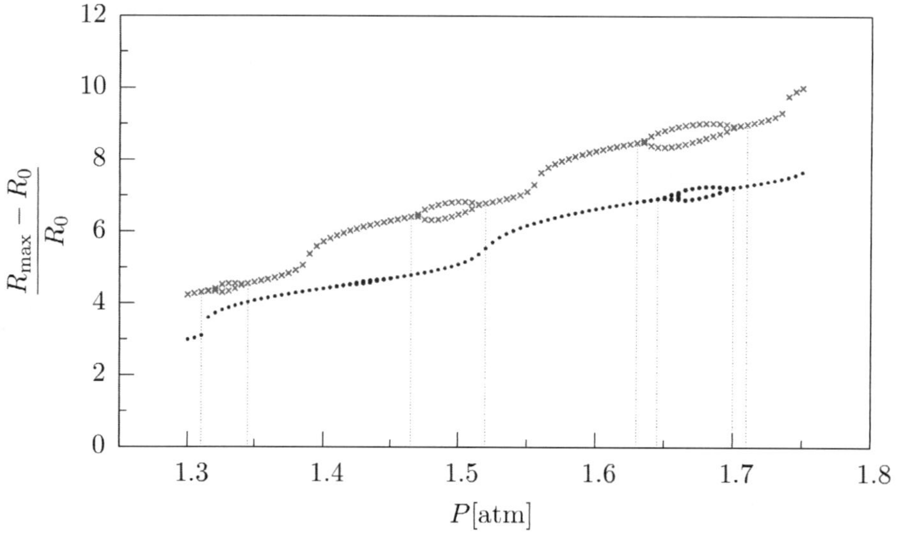

(a)

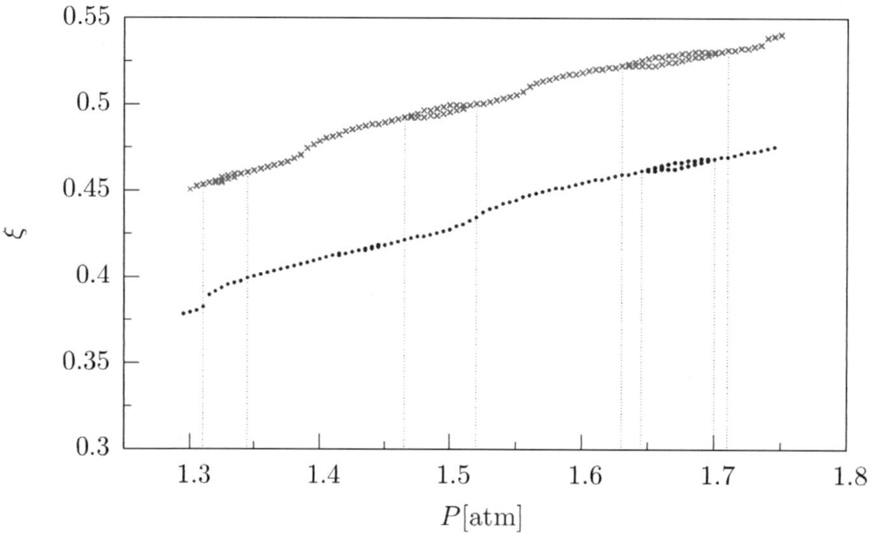

(b)

Figure 3. Bifurcation diagrams for a bubble determined by the Rayleigh-Plesset equation (9) with parameters given in figure 2 but with initial radius $R_0 = 0.9\mu$m. In both subfigures data from the fifty cycles is omitted in order to disregard transient behaviour. The radii from linear wave propagation are shown as red crosses, those from nonlinear wave propagation shown in blue dots. Subfigure 3(a) shows the change in the maximum amplitude, whereas subfigure 3(b) shows the change in the phase of the collapse, but the same qualitative features are displayed, that is period doubling bifurcations occur for lower pressures for linear waves than nonlinear waves.

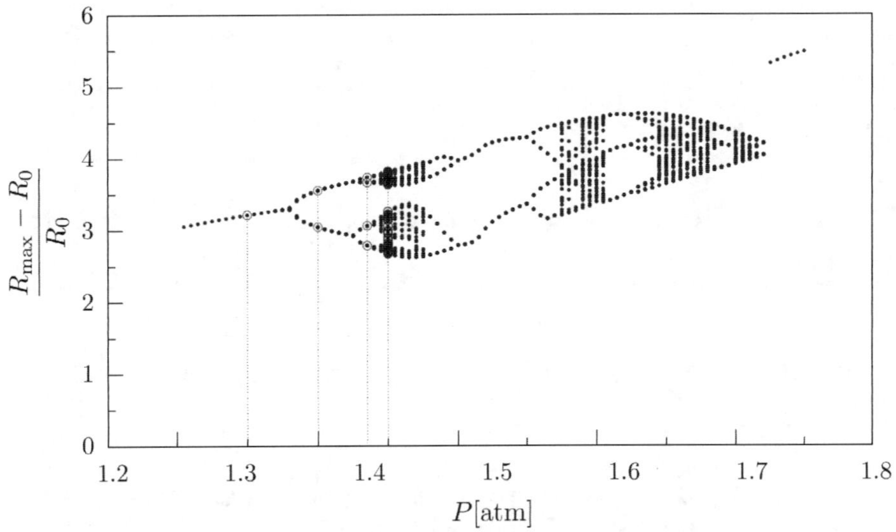

Figure 4. Bifurcation diagram showing the period doubling and the onset of quasi-periodic oscillations as the magnitude of the forcing pressure of the Fubini solution (5) is varied for the Rayleigh-Plesset equation (9). The parameters are the same as figure 3 except that now $R_0 = 1.4\mu$m. Profiles of the radius time curves at the marked pressures are illustrated in figure 5.

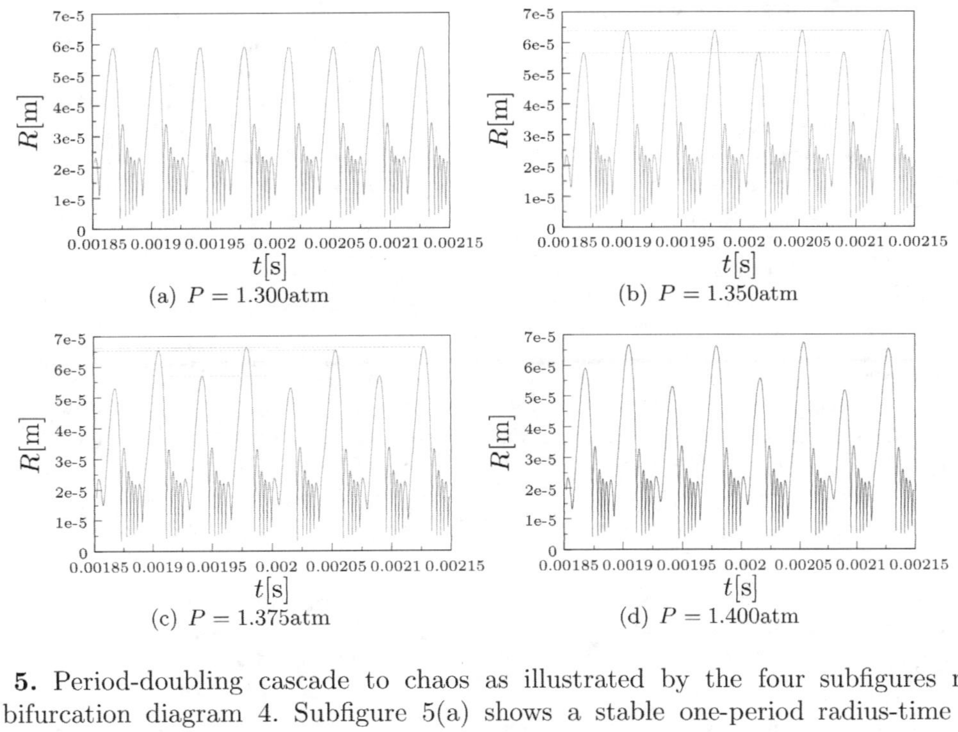

Figure 5. Period-doubling cascade to chaos as illustrated by the four subfigures marked on the bifurcation diagram 4. Subfigure 5(a) shows a stable one-period radius-time profile at $P = 1.30$atm, which undergoes a period doubling bifurcation so that at $P = 1.35$atm in subfigure 5(b) the profile is two-period. For $P = 1.375$atm as subfigure 5(c) is four period before subfigure 5(d) shows a chaotic profile at $P = 1.40$atm.

Anglo–French Physical Acoustics Conference 2009 IOP Publishing
Journal of Physics: Conference Series **195** (2009) 012009 doi:10.1088/1742-6596/195/1/012009

Transient space-time surface waves characterization using Gabor analysis

L Martinez[1], N. Wilkie-Chancellier[1,3], C. Glorieux[2], B. Sarens[2] and E. Caplain[1]

[1] Université de Cergy Pontoise, ENS Cachan, UMR CNRS 8029, Laboratoire Systèmes et Applications des Techniques de l'Information et de l'Energie (SATIE), 5 mail Gay-Lussac, F 9500 Cergy-Pontoise, France
[2] Katholieke Universiteit Leuven, Laboratorium voor Akoestiek en Thermische Fysica (LATF), Celestijnenlaan 200D, B-3001 Leuven, Belgium

E-mail: nicolas.wilkie-chancellier@u-cergy.fr

Abstract. Laser ultrasonics allow the observation of transient surface waves along their propagation media and their interaction with encountered objects like cracks, holes, borders. In order to characterize and localize these transient aspects in the Space-Time-Wave number-Frequency domains, the 1D, 2D and 3D Gabor transforms are presented. The Gabor transform enables the identification of several properties of the local wavefronts such as their shape, wavelength, frequency, attenuation, group velocity and the full conversion sequence along propagation. The ability of local properties identification by Gabor transform is illustrated by two experimental studies: Lamb waves generated by an annular source on a circular quartz and Lamb wave interaction with a fluid droplet. In both cases, results obtained with Gabor transform enable ones to identify the observed local waves.

1. Introduction

Laser detection methods allow the investigation of ultrasonic transient phenomena in both space and time dimensions. Used along a two dimensional surface, laser ultrasonic leads to three dimensional (3D) space–time signal collections. Straight 3D Fourier analysis or classical high resolution signal processing methods can be used to identify the wave propagation information, however these methods are not adapted for identifying where and when the waves are generated. The 3D Gabor transform is presented in the view to quantify these transient aspects in the space–time–wave numbers–frequency domains. Two experimental studies illustrate the ability of local properties identification by Gabor transform, the first one in the case of Lamb waves generated by an annular source on a circular quartz sensor and the second one for Lamb wave interaction with a fluid droplet.

[3] To whom any correspondence should be addressed.

2. 3D Gabor transform

The Gabor transform was introduced in 1947 [1] in order to achieve a signal analysis closer to human perception of sound such as time localization of frequency components. It gave the principle of the time–frequency analysis by short-time Fourier transform (STFT), with the help of a sliding gaussian window [2-4].

The early applications dealt with audio signals and some efficient compression methods rose from the comprehension of their frequencies components time shape. The 2D Gabor transform were introduced in acoustics [5] in order to localize ultrasonic phonons in space–time domains and their conversions on a cylindrical shell. The 3D Gabor analysis presented here is both an extension to three dimensions and an improvement of the early 2D version cited above [6].

2.1. Localisation of phonons by 3D Gabor transform

The goal is to estimate the local space and time periods from propapagating waves signals and to map their amplitudes through the space–time domain. The fundamental hypothesis is that the propagating waves can be observed along a 2D surface on points referenced by their coordinates (x,y) at time t. This leads to the 3D space–time collection of the observed waves s(x, y, t). Along an infinite space and time field of view (FOW), by following the Fourier diamond [7-13], three other spaces can be deduced from the original signals s(x,y, t) (Figure 1 left) by Fourier transforming s(x,y,t):

- S(x,y,ω) is the time Fourier transform of s(x,y, t),
- N(kx,ky,t) is the space 2D Fourier transform of s(x,y,t) and
- Ksi(kx,ky,ω) is the 3D Fourier transform of s(x,y,t) along the space and time dimensions.

The 3D Gabor transform uses a small space and time cube sliding FOW prior Fourier transforming (Figure 1 right). The dimensions of these cubes will be chosen small enough in order to not overlap two successive wave packets (i.e. keep space-time resolution) and wide enough to keep enough resolution in the frequencies domains. A compromise must be reached but the sliding cube has to cover at least one wavelength in the spatial dimensions and at least one period in the temporal dimension.

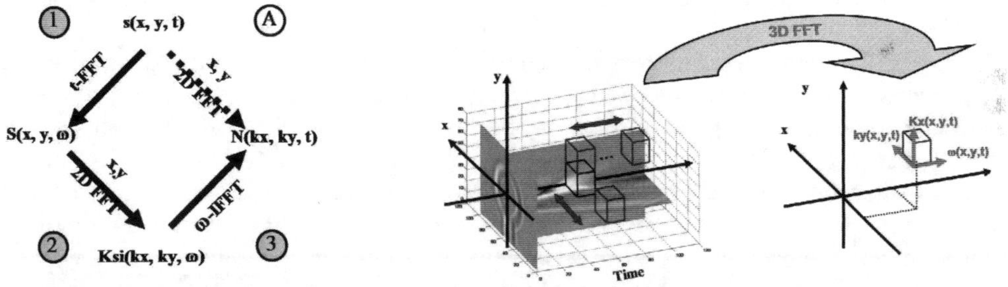

Figure 1. Fourier diamond (left) and 3D Gabor transform principle (right).

For 2D signals s(x,t), these 3 spaces have interesting properties for the modal analysis of the observed waves. The 2D Fourier transform of s(x,t) can be used to quantify, a posteriori, the wave properties, like their attenuation and phase velocities [6–13]. With the help of Prony identification, the mixed representations S(x,ω) and N(kx,t) can be also used successfully [14-15]. However when the signal s(x,t) is a sum of identical wide band wave packets delayed in time and in space, identification based on S(x,ω) allows ones to precise where the waves are packets, not when, whether N(kx,t) allows the time localization, not the space one, and Ksi(kx,ky,ω) none of both as everything is accumulated.

For 3D data signals s(x,y,t), some of the 2D tools are still usable: S(x,y,ω) is useful for identifying the wave modes. However, the problem of wide band wave localization in both time and space encountered for the 2D case remains in the 3D case, and a second difficulty appears: identification by Prony like methods is not possible without a deeper prior knowledge of the experimental signal composition [14-15].

2.2. 3D Gabor scheme

The 3D Gabor transform Z(x,y,t,kx,ky,ω) can be obtained through several equivalent paths. The first possible method is the straightforward extension of the time frequency methods STFT to three dimensional signals: a small 3D cube slides along the time and space dimensions, and the corresponding 3D spectrums Ksi(kx,ky,ω) are stacked, leading to the Z(x,y,t,kx,ky,ω) representation. However, as Z is a 6D space of complex numbers, this method needs a huge amount of computer resources to be computed and stored. Moreover the resulting Z(x,y,t,kx,ky,ω) representation is not easily readable from a physical point of view.

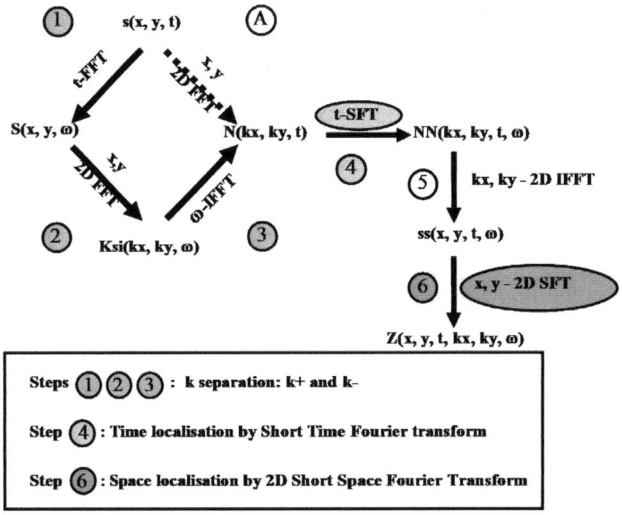

Figure 2. 3D Gabor analysis scheme and connections with the Fourier diamond spaces.

In order to reduce the computer resources needed to reach the 6D Z and to give Z a physical meaning the x, y, t localisation is done in successive steps instead of a huge one (Figure 2). The proposed scheme presented here uses the physical fact that energy flows, through the time dimension, from past to future, whereas, through the space dimensions, both directions are used. As the s(x,y,t) collection is real valued, loosing the negative frequencies that correspond to propagation towards negative time, one can orient the space propagation and separate positive wave numbers from negative ones. This is the goal of following the path (1)–(3), instead of the path (A) in Figure 3. The obtained wave number–time representation N(kx,ky,t) is then wave number oriented and the waves propagating along increasing x or y directions are separated from the one propagating in the opposite directions. The short time sliding Fourier transform (t-SFT) is then applied to N(kx,ky, t) along the time dimension (Step 4). The resulting 4D NN(kx,ky,t,ω) representation allows the localization of the wave number–frequency (kx,ky,ω) aspect through time, but still not through space. The energy localisation through space is then done by following the next two steps:

(5) NN(kx,ky,t,ω) is sent back to the space dimension by inverse space Fourier transform, leading to an intermediate space–time–frequency representation ss(x,y,t,ω),

(6) the Z(x,y,t,kx,ky,ω) representation is the short space sliding Fourier transform (x,y-2D SFT) of the ss(x,y,t,ω) representation. Instead of storing the complex values Z(x,y,t,kx,ky,ω), the wave numbers kx and ky are estimated from an identification among Z, by identifying each modulus maxima. Complex wave numbers can be extrated from the complex representation Z : the kx and ky real parts locate the peaks in Z, whereas their imaginary parts shape the peaks height and width. The real part of the wave numbers corresponds to the propagation, whereas their imaginary part represent

the attenuation coming from both the propagation and the geometry of the observed wave source (such as plane, circular). In this paper, we chose to only scan the propagation given by the real part of wave numbers. Choosing a frequency of interest x enables us to image the space–time evolution by superimposing the real part of identified wave numbers on video sequences of ss(x,y,t,ω).

3. Quartz sensor disk transients

The first application of the 3D Gabor analysis is the identification of the waves generated by an electrical pulse on a quartz sensor surface.

3.1. Quartz sensor and experimental set-up

The quartz used is a standard AT cut disk (thickness 0.3 mm, diameter 15 mm) designed to have a 6 MHz fundamental frequency (figure 3 left). The deposited gold electrodes are 5 mm in diameter. The quartz is maintained only by the electrodes and a thin chromium layer is deposited in order to impose the detection of vibration along the upper face of the quartz, with the same sensitivity, whether the laser beam is on or outside the upper gold electrode.

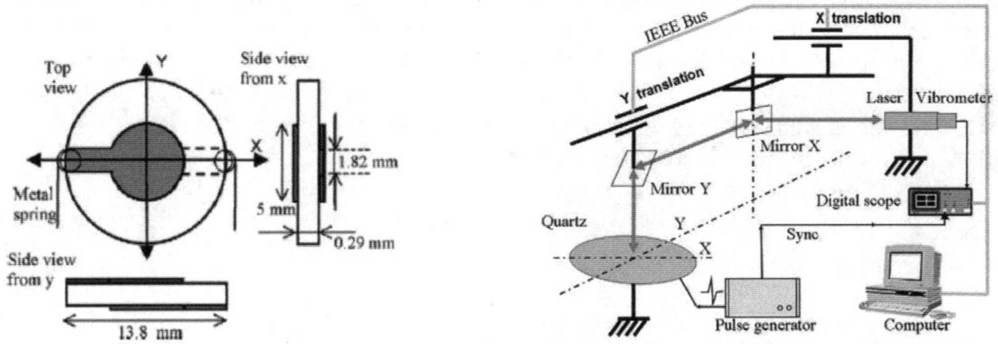

Figure 3. Quartz sensor (left) and the experimental set-up used (right)

The quartz is excited by a voltage pulse (amplitude 200 V, bandwidth 100 MHz) in order to generate waves in a wide frequency range (figure 3 right). A Polytec laser vibrometer is used for the optical detection of the acoustic waves propagating along the quartz surface. Due to the single head used, only mainly normal vibrations are measured on the quartz surface. A computer is used for both the time signal acquisition through a Lecroy digital scope and for the positioning of the laser beam through automated micro-translations.

The space acquisition is performed over rectangular area that nearly included the whole area of the quartz (80 by 110 positions, with 125 μm steps). The frequency bandwidth used is 50 MHz.

3.2. Signals and 3D Gabor analysis

Figure 4 (left) presents the space-time signals obtained by the optical detection on the upper surface of the quartz.

The snapshots in Figure 4 (right) present slices of the space-time-frequency representation ss(x,y,t,ω) for a selected frequency F = 885 kHz. The arrows lengths patched on the images are proportional to the real part, i.e. the propagating part, of the local wave numbers estimated from the 3D Gabor transform. The arrows point out the corresponding local direction of propagation.

The equivalent sliding cube used is 16x16 points along the space dimensions and 64 points along the time dimension. These values have been selected small enough in order to not overlap two successive wave packets and wide enough to keep enough resolution in the frequencies domains. In this case, the sliding cube covers one wavelength in the spatial dimensions and one period in the temporal one.

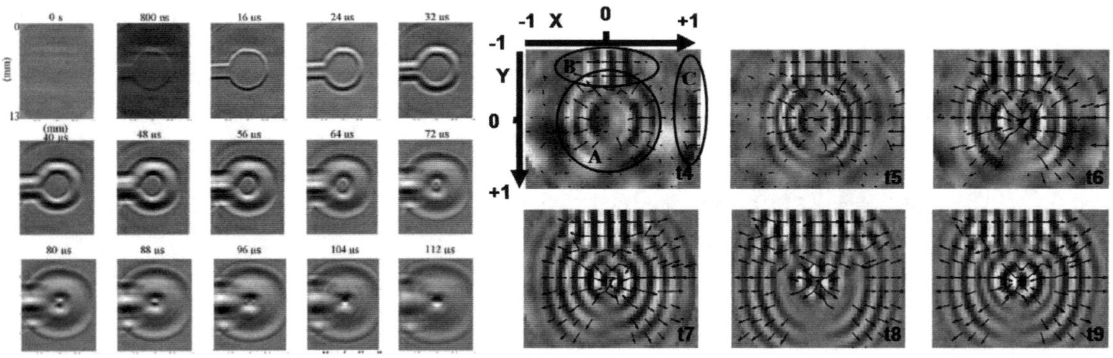

Figure 4. Space-time signals (left) and Gabor analysis at F = 885 kHz (right).

The estimated wave numbers directions confirm raw observations made from the dataset.
In zone A and B (figure 4 right), the opposite wave numbers are clearly identified for the waves generated by the electrode border. In zone A (respectively B), the circular (respectively linear) shape of the wave front is clearly shown with the radial wave numbers. For both zones, the corresponding phase velocity is close to 1770 m/s, in good agreement with experimental results presented in [16], and obtained by modal analysis. This corresponds to the A_0 Lamb mode, dispersive in this frequency domain. There is also a good agreement with the recent theoretical results presented in [17] for the flexural mode F, dispersive in this frequency domain.
In zone C, one can notice that waves are generated by the border of the quartz.
For other frequencies, A_1, S_1 and S_2 waves are identified, as their space and time generation points (Figure 5).

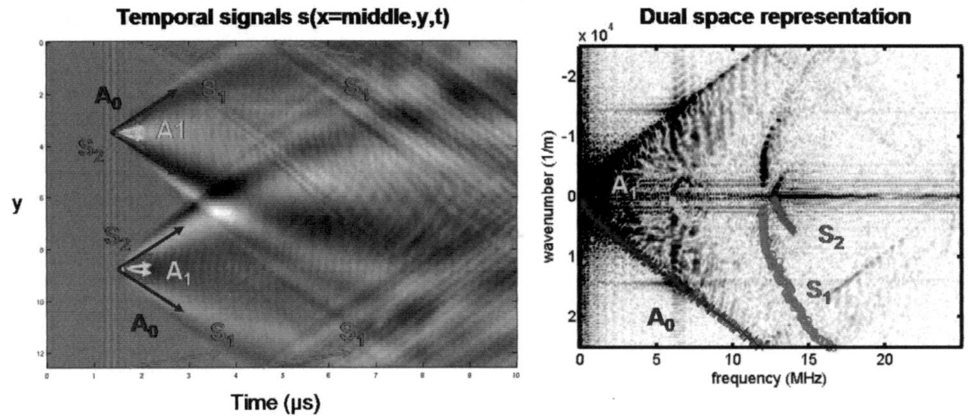

Figure 5. Identified waves for all frequencies in the spatio-temporal (left) and dual spaces (right).

4. Droplet interactions with Lamb modes

The second illustration of the 3D Gabor analysis is the interaction of Lamb waves with a fluid droplet.

4.1. Experimental set-up

In this case, a 20 μL glycerol droplet (diameter 6 mm) is put on the surface of an aluminium plate. The generation of Lamb wave is performed in pulse mode using a 2.25 MHz transducer in contact with the edge of the plate. The laser vibrometer scans a 20x20mm area surrounding the droplet, by 0.5mm steps. A computer is used for both the time signal acquisition through the Lecroy digital scope and for positioning the laser beam through automated micro-translations.

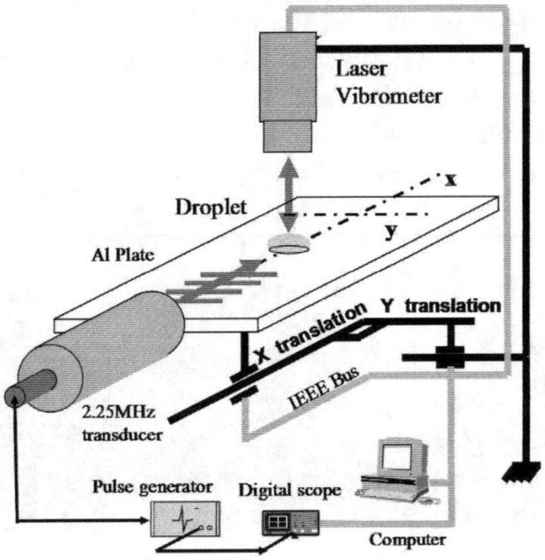

Figure 6. Experimental set-up for the study of the droplet behavior.

4.2. Signals and 3D Gabor analysis

Figure 7 (left) presents the space-time signals recorded by the optical detection on the upper surface of the aluminium plate, along the area surrounding the droplet. The interaction of the surface waves with the fluid droplet can be clearly observed.

From 3D Gabor analysis and for a selected frequency (F=1MHz), the spatio-temporal evolution of the displacements can be represented at y=0 (Figure 7 right), i.e. a cut of the previous representation. Several observations can be done. In a first hand, two different incident Lamb waves are visualized. The first generated mode is the S_0 Lamb mode and the other one is the A_0 Lamb mode. On the other hand, the interactions of the S_0 and the A_0 Lamb modes with the droplet are also observed successively.

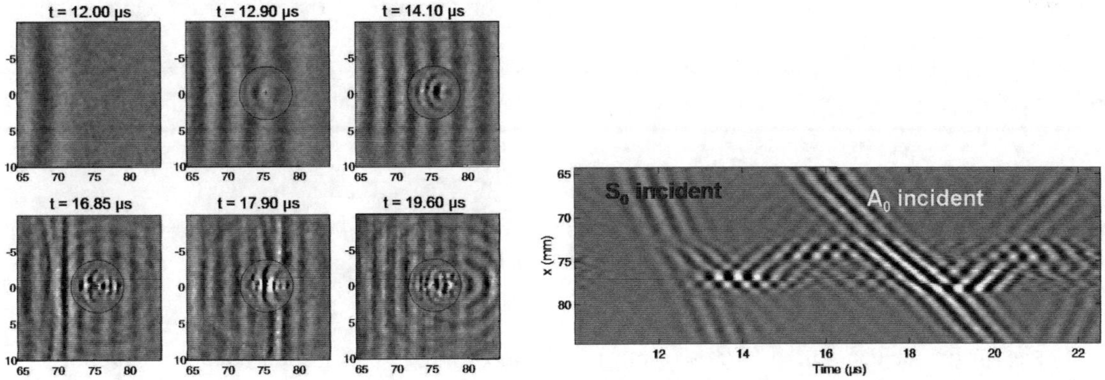

Figure 7. Space-time signals (left) and representation at y=0 (right) for F=1MHz.

The snapshots in Figure 8 are space-time-frequency representation Z(x,y,t,kx,ky,ω) for a selected frequency F = 1MHz and at several times. So, for each picture, the (kx,ky) informations are superimposed on the 2D spatial evolution s(x,y). Using this point of view, the local wavenumbers are identified [18].

On the three first snapshots, the interaction of the S_0 Lamb mode with the gycerol droplet is observed. The incident S_0 Lamb mode (t = 12.00 μs) travels under the droplet and is converted into A_0 Lamb mode and A mode at t = 12.90 μs, along the droplet border. As the A mode is a propagating mode which is characteristic of the fluid constituting the droplet, it is trapped inside and travels back and forth. The droplet focusing effect is clearly visualized on the second snapshot.The A_0 and A converted modes are propagated starting from this focal point (t = 14.10 μs).
Using the three next snapshots, the interaction of the A_0 Lamb mode with the fluid droplet can be understood. The second incident Lamb mode (A_0) is incident and converted into A mode. A second focal point can be noted starting from which the A_0 and A modes are propagated (t = 19.60 μs).

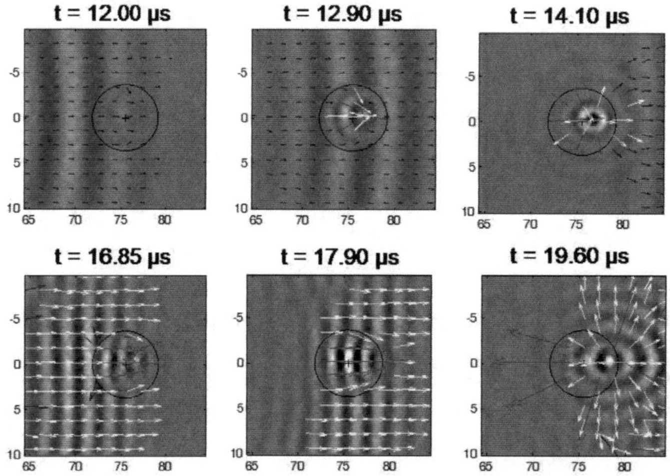

Figure 8. Space-time signals and Gabor analysis for the droplet interaction with Lamb modes at F = 1 MHz (S_0 mode in blue, A_0 mode in yellow and A mode in red)

We can note that the positions of the two focal points are not similar for t = 14.10 μs and t = 19.6 μs. This is due to the difference of velocity between the A_0 and the S_0 Lamb modes, as the refraction of the converted waves follows the Snell's law. A change of the incident wave velocity gives rise to a change of the index and then induces a change of the convergence of the converted wave along the curved droplet boundaries.
The experimental determination of the focal point position could give access to the velocity of the A mode in the droplet, and then could allow us to characterize the material constituting the droplet.

5. Conclusion

The 3D Gabor transform has been successfully employed for identifying the local wave numbers and frequencies present in transient 3D signals. The experimental results are in good agreement with theoretical modes predictions for an infinite plate.
Moreover, in the case of the quartz sensor behavior, the role played by the shape of the electrodes is clearly shown for the generation of these modes. This knowledge enables theoretical modal analysis of the real quartz sensor and to tune its electrodes shape.
In the case of the droplet study, the conversion of Lamb modes during the interaction of an incident wave with the fluid droplet allows us to monitor the droplet behavior.
Both cases pointed out that 3D Gabor transform is well adapted to the analysis of transient waves propagating with various wave fronts shapes, converting in other waves. Further investigations on non linear wave propagation cases should also give useful results.

6. References

[1] Gabor D., 1946 , Theory of communication, *Proc. Inst. Electric. Eng.*, **93 (26)** 429–457.

[2] Claasen T.A.C.M, Mecklenbraüker W.F.G, 1980, The Wigner distribution. A tool for time-frequency analysis, Part I: Continuous-time signals, *Phillips journal of research*, **35 (3)** 217-250.

[3] Claasen T.A.C.M, Mecklenbraüker W.F.G, 1980, The Wigner distribution. A tool for time-frequency analysis, Part II: Discrete-time signals, *Phillips journal of research*, **35 (3)** 276-300.

[4] Claasen T.A.C.M, Mecklenbraüker W.F.G, 1980, The Wigner distribution. A tool for time-frequency analysis, Part III: Relations with other time frequency Continuous-time signal representations, *Phillips journal of research*, **35 (6)** 373-389.

[5] Martinez L., Morvan B., Izbicki J.L., 2004, Space-time-wave number-frequency X(x,t,k,f) analysis of SAW generation on fluid filled cylindrical shells., *Ultrasonics International*, **42 (1)** 383-389.

[6] Martinez L., Goossens J., Glorieux C., Wilkie-Chancellier N., Ould Ehssein C., Serfaty S., 2006, 3D Gabor analysis of transient waves propagating along an AT cut quartz disk, *Ultrasonics International*, **43** 1173-1177.

[7] Alleyne D., Cawley P., 1991, A two-dimensional Fourier transform method for the measurement of propagating multimode signals, *J. Acoust. Soc. Am.*, **89 (3)** 1159-1168.

[8] Vollmann J., Dual J., 1997, High-resolution analysis of the complex wave spectrum in a cylindrical shell containing a viscoelastic medium. Part I. Theory and numerical results, *J. Acoust. Soc. Am.*, **102** 896-908.

[9] Vollmann J., Breu R., Dual J., 1997, High-resolution analysis of the complex wave spectrum in a cylindrical shell containing a viscoelastic medium. Part II. Experimental results versus theory, *J. Acoust. Soc. Am.*, **102** 909-920.

[10] Salisbury J.I., 1999, Wave number frequency (k-ω) analysis using wavelet transform and eigenvalue decomposition, *J. Acoust. Soc. Am.*, **106 (3)** 1602–1604.

[11] Bonnet G., 1983, Au delà d'une vitesse de groupe: vitesse d'onde et vitesse de signal. Première partie: l'opérateur vitesse de groupe en l'absence d'affaiblissement, *Ann. Télécommun.*, **38 (9-10)** 1-22.

[12] Bonnet G., 1983, Au delà d'une vitesse de groupe: vitesse d'onde et vitesse de signal. Deuxième partie: déformation de l'amplitude et influence de l'affaiblissement, *Ann. Télécommun.*, **38 (11-12)**, 1-17.

[13] Martinez L., 1998, Nouvelles méthodes d'identification d'ondes de surface. Etude de l'onde A sur une cible courbe, *PhD Thesis*, Université du Havre, France.

[14] Martinez L., Duclos J., Tinel A., 1998, Aspect of cylindrical shell resonances in Fourier Diamond Spaces. Use of Surface Waves Analysis Methods (SWAM) on experimental or numerical datas, *J. Acoust. Soc. Am.*, **103** 2901.

[15] Martinez L., Morvan B., Izbicki J.L., 2003, Phonon localisation methods through time and space: experimental SAW phonon aspects on a cylindrical shell, *J. Acoust. Soc. Am.*, **113** 2294.

[16] Leclaire P., Goossens J., Martinez L., Wilkie-Chancellier N., Serfaty S., Glorieux C., 2006, Study of the bending modes in circular quartz resonators, *IEEE Trans. UFFC*, **53(10)**, 1934-1943.

[17] Lee P.C.Y., Liu N., 2004, Plane harmonic waves in an infinite piezoelectric plate with dissipation, *IEEE Trans. UFFC*, **51 (12)**, 1629–1638.

[18] Martinez L., Wilkie-Chancellier N., Glorieux C., 2008, 3D Gabor analysis of transient space-time waves, *J. Acoust. Soc. Am.*, **123 (5)** 3432.

AUTHOR INDEX